THE INNER CLIMATE

Global Warming from the Inside Out

BY THEO HORESH

For everyone working to stop climate change in their own special way.

Library of Congress Cataloging-in-Publication Data

Horesh, Theo

The Inner Climate:
Global Warming from the Inside Out

p.cm

1. Environment. 2. Climate Change. 3. Global Warming.

ISBN 13: 978-1-936955-17-6

Bäuu Press
Golden Colorado
www.bauuinstitute.com

CONTENTS

INTRODUCTION

We cannot see climate change, but its results are everywhere. We cannot hear it coming, but the things we hear are to come do not sound good. Just about everything we do in life contributes to climate change, and yet we can identify no clear victims of our actions. There are crimes but never a smoking gun; there are harms but the impact is diffuse and responsibility shared by billions. We are told to be concerned for our children's futures, but we now know that climate change will affect lives thousands of years in the future, so messages like these can sound like propaganda. We are told to drive less and eat less meat, and yet we know such injunctions have barely dented behaviors and that much collective action will be necessary. While most Americans support doing something about climate change, and while many believe it could destroy civilization, climate change tends to score last on the list of political priorities. It is as if there is something about climate change about which we cannot make sense. If only climate change had been started by Al-Qaeda or President Obama had tried to cover it up, the demand for action might be more forthcoming.

Climate change is an emergency the nature of which has hardly changed in a generation. The basic science has not changed; the nature of the solutions remain the same. It is like one of those films in which the few seconds left to save the world are stretched across the last several minutes of the closing scene. Those of us who track it closely can vacillate between sitting back and waiting for the clean energy revolution and thinking the world is about to end. Climate change is occurring on such a grand scale that it confounds our minds. If we are to understand its impact, we must stretch our minds spatially to the ends of the earth, temporally to all future generations, and categorically to all life on the planet. Having made the necessary changes to our mental frameworks, we must somehow take responsibility for every living being unto the end of times. But this is too much to ask from even a saint. Thinking about climate change can be overwhelming.

Humans evolved to notice immediate changes in our near vicinity, to wake to the leopard or the slithering snake. We did not evolve to notice gradual developments, like the rise in childhood obesity. We tend to worry not about those risks that are the greatest threat; rather, we worry about risks most salient in our minds, the ones we can imagine and easily identify. We can easily identify childhood obesity, picture its victims, and feel for their plight. But climate change is more an invisible killer, who strikes when we are unaware and sometimes plays tricks on our minds. He makes us think he is here in the hurricane when really he is lurking 'round the corner in the melting glaciers. Thinking about climate change can be confounding.

To reason about climate change we must be able to think globally. This may not be as easy as it sounds. The world is vast and involves more variables than say, fixing the debt or reforming health care. There are more people involved and more contingencies to consider. Tracking more variables requires greater study. And it expands the range of what we deem to be important. It is hard to think globally if we cannot transcend our American or British or Indian concerns. But such a process can take decades, and it can be tumultuous. Group identities run so deep that renouncing them often means building new identities, and this takes time. It is a complex and often emotional process, a bit like differentiating yourself from your parents. Perhaps there are ways to think more globally that bypass this process – perhaps we can pass into adulthood without teenage rebellions - but we need to notice how these developments do happen. Thinking about climate change is time consuming.

Climate activists are in a particularly tight spot. Since people may not be prepared to hear what they have to say, they must find ways of making their message vivid. Drowning polar bears and flooded cities can do the trick. But they are not really our greatest worries. Still, there is a bigger challenge. Climate science looks at probabilities, activists at eventualities. The sense that something will happen unless we act brings people together to shape their shared fates. This is hard to do when speaking of probabilities. When we hear that an event is "somewhat likely," and dependent on a complex set of variables, the probability we will act is sadly somewhat unlikely as well. The science also distinguishes probabilities from possibilities. It is possible climate change will destroy civilization, but it is not probable. And there is a dishonesty involved in making an outcome for which there is a 5 percent chance appear inevitable. Activists are in a sort of double-bind. If only people understood what was really happening they would act, but it is all too nebulous. So, activists simplify their messages by distorting the science. But this downgrades the science, and climate skeptics close in like wolves. Thinking about climate change can be paralyzing.

Climate change challenges us to care for the whole of the world. But humanity evolved to care for much smaller groups. Most of us were conditioned to think of our own nations. We know how to think about a state, how to reason about a new social program or military venture. And most of us know how one issue effects another in our own states. Social programs and military ventures come at a price, but sometimes we can expand the pie. If we are adults, we may have an idea how. This is elementary national politics. But few of us can make much sense of the trade-offs involved in ending hunger and stopping climate change, halting deforestation and protecting the oceans. They may be all of a piece but this is a puzzle for which we have not put in the time. Climate change challenges us to weigh these issues against each other, to pick winners and set priorities. This can be contentious. And it is a process for which few have training. Thinking about climate change can be complex.

Humanity has confronted losing the world through nuclear war, but never have we confronted so threatening a peril as climate change for which we might all be held responsible. Climate change is largely the result of the human desire for greater comfort. The places we live, the things we eat, the way we get around: these lie at the heart of what will need to change. So, climate activists can sometimes sound as if they want us to renounce everything pleasant. We are often told that our Priuses and those nice new solar panels will not even make a difference. So, we can start to feel as if we are killing the world softly and that there is nothing we can do about it. Thinking about climate change can be heart-breaking.

Conservatives have had a particularly hard time with climate change. Most conservatives seem to be concrete thinkers. They are most likely to be moved to action by the things they have seen and felt. But climate change is an abstraction whose results tend to be measured in probabilities. Whether they are hard nosed realists or cautious moderates, conservatives also tend to be skeptical of the collective changes that may be involved in solving climate change. When climate activists put a positive spin on the changes that need to occur, their visions can involve a wish list of progressive programs. So, it has been a challenge to get conservatives on the bus. On top of all that, conservative politicians tend to have stronger ties to oil and coal; theirs tends to be a dirtier bus. And publicly shaming them about it has often come at the price of their involvement. There are many fights that need to happen if climate change is to be taken seriously, but fights create enemies and enmity harms our ability to work together. Thinking about climate change can be contentious.

But climate change has not always fit well into a progressive agenda either. Since greenhouse gas emissions tend to rise as economies grow, climate activists often oppose development. Economic growth is more often than not treated as an enemy. And this has not helped win support amongst the poor of the developed or the developing world. Climate activists often find themselves defending the needs of the yet unborn against the aspirations of the living. Since future generations do not vote in the elections of today, climate ethics have not fit well into winning political platforms. Hence the turn to personal ethics and local solutions. But as we have seen, such solutions have been said to be no solutions. So, climate activists can talk as if they wish things would get worse before they get better. For then we might wake up and act. Needless to say, this is a dangerous approach. Thinking on climate change can be self-defeating.

Overwhelming and confounding, self-defeating and paralyzing, contentious and heart-breaking: thinking about climate change can be tough. It is a big topic, with numerous components, conflicting evidence, competing solutions, lots of controversy, serious consequences, and a relatively small constituency. Thinking about climate change pushes our minds to the limit. And it often involves walking through emotional minefields. We are pressed to be neutral and objective when we talk about the discontinuation of human civilization. We are asked to be civil when confronting lies with epic consequences. World leaders might learn to think like this in the midst of great wars, but they are usually selected and trained for the task. And they tend to have the smartest advisors to aid in their prudence. But now everyday people are enjoined to reason through the future of life on earth. We are enjoined to save the world – as if we even knew how to save ourselves. But perhaps we can see the problem in a different light.

Thinking about climate change can be stimulating and novel, enriching and inspiring, rewarding and expansive, and even sublime. These are not the sorts of feelings we commonly associate with thinking about anything. But climate change is different from other mental challenges. It can turn our worlds inside out and open up new trajectories. Mike Hulme suggests we should, "ask not what we can do for climate change but what climate change can do for us." For climate change brings everything into question. And in so doing, it allows us each and all of us together to make a fresh start. Climate change draws so many aspects of our civilization into question that it challenges us to think again about how we live together. As Peter Senge suggests, climate change is like getting cancer at 40. It forces humanity to rethink the way we live. Thinking about climate change needs to be embraced as an opportunity.

It is into this thicket of mental and emotional challenges that *The Inner Climate* is thrust. *The Inner Climate* explores how to grapple with climate change within and among ourselves. It is difficult to know how to respond appropriately to such an amorphous and paralyzing threat as climate change. So, one might expect climate thinkers to devote some attention to the inner climate of how we think and feel about the challenge. Many thinkers do tell us how we should think and feel about climate change, and the answers help. Some of the best answers have originated with many of this book's contributors. But seldom do we reason through the multitude of often contradictory programs and inquire into which responses are best. Rather we tend to take our reactions to climate change as a given, like an adolescent lover swept up in a fight. Thinking about climate change needs to be more reflective.

The human response to climate change has largely been neglected, and many in the climate community have begun to see this as a deficit. Few authors touch on the ambiguities involved in doing probabilistic science. And few say much about the emotions evoked by these ambiguities. Climate change might make the planet unlivable, but it might also be quickly resolved through a series of technological solutions. And it might be partially solved in a way that causes significant hardship and economic loss. Such uncertainties can evoke neurotic responses. Since we cannot tell whether the world is ending or whether it is no big deal, some people freeze up, some become frantic, some get resistant, some fatalistic. These emotions can get in the way and hinder action. And they can lead us to look askance. There are only so many intense challenges each of us is willing to consider, after all, and many of us already have a full plate. Making writings on climate change more palatable can increase the demand for them. And researchers and organizations whose inner climates are more temperate can produce much more appealing writings.

There is another problem with writing on climate change. Few thinkers writing on the issue seem open to a free exchange of ideas. This is not a field like international relations in which good people can differ graciously over priorities and responsibilities. Rather, climate change is more like a battle in which each side throws everything they have got against the other side. Such battles tend to demand full allegiance. Challenging questions can put a stop to campaigns. They can sidetrack attention and breed dissent. And they often get exploited by the opposition. Climate activists are besieged by doubters whose motives are questionable. So, there is a real need to get on with business. But there are also real disagreements among climate activists. There is debate over whether action should be personal or political, local or global. There is renewable energy versus nuclear power, intensive agriculture versus small-scale organics, genetic engineering versus the precautionary principle. These debates can hinge on technical questions. But they also involve values and visions of the good life. Our answers depend on personal preferences, but there is a tendency to act as if more data will simply solve the problem. Insofar as climate change brings everything into question, it challenges us to question everything. And this is difficult to do in the midst of a campaign. Few great works of philosophy have been written in the midst of campaigns. But people do tend to turn philosophical when everything is up for question.

There is an inner climate to the climate movement itself. The tone of disagreement, the feelings evoked, and the manner of reasoning, all impact the nature of the climate movement. Whether or not people will listen is often dependent on the psychological health and the emotional maturity of the speaker. Whether or not they act can depend on their own inner dispositions. Even if we stop adding CO_2 to the atmosphere tomorrow, the world will still keep getting warmer. So, we do not only need to develop sustainable practices, we need to develop a sustainable relationships to climate change, because it is a reality we must confront for the rest of our lives. The climate movement would do well to take its own sustainability more seriously.

There is an inner climate to the climate science as well. Personal values and styles of thinking effect the questions scientists take up, the way they work together, the models they use, how they carry out their research, the way it is conveyed to the public, and the level of trust they are granted by other experts and members of the public. Scientists are human, and at each stage of their work there are choices involved. Scientists can never be fully neutral, if only because even the decision to try to be neutral involves subjective values concerning the nature and purpose of science. The decision to take up a given science in the first place is also often motivated by a concern for the results the science might yield. None of this need imply that science is hopelessly biased. In fact, the awareness of biases allows researchers to account for them and to build ever more objective models. But paying greater heed to the subjective side of science highlights the moral responsibility of scientists to share what they know, to clear up misunderstandings, to bring attention to its import, and perhaps even to sound the alarms. We would expect nothing less from a Russian military historian, after all, who saw all the signs of imperial expansion where others saw only domestic turmoil. But if such a historian did ring the warning bells, we would also hope that he gracefully respected the contributions of military strategists, political leaders, and nuclear disarmament specialists to the debate.

This book draws on a colorful spectrum of thinkers. An ethicist, a journalist, a geographer, and a psychological researcher; a linguist, an economist, an ecologist, and a sociologist; a meteorologist, an anthropologist, an epidemiologist, and a management theorist. As the introductory bios will make clear, the contributors to this volume are often the defining thinkers in their respective disciplines. Yet, most are also interdisciplinary thinkers and many resist easy definition. Each of them has something special to share about our own inner climates, but sometimes getting there involves talking in a roundabout manner. Climate change is a big issue that must be confronted in multiple domains. It cannot be left to the climate scientists and economists, the politicians and activists. It is a big enough challenge to include us all, so big that understanding reactions to it can involve touching on a wide range of human knowledge. In this light, we might say that each of these thinkers holds a vital piece of the puzzle concerning how we think about the inner climate. But readers should not expect to come away with a complete picture.

So, perhaps we should think of this book as being more akin to a collage, whose pieces are overlapping, interrelated, and associative. Both its beauty and its usefulness lie in its diversity. If this book is a success, it will raise as many questions as it answers. And it will bring to light multiple pathways out of the morass. This approach may be controversial to activists, who often seek to narrow the terms of discussion in an effort to prevent manipulation of the debate and the evasion of difficult choices. But a basic premise of this book is that there are simply too many things that need to be done, too many people that need to be brought on board, too many adaptations that need to be made, for any but a broad and inclusive dialogue. This does not mean we should not be talking about a global climate deal or a national carbon tax. Rather, even these sorts of comprehensive solutions are not enough and are, in actuality, usually just overarching ways to incentivize a multitude of other more local changes that will ultimately need to occur.

This is not the first book you should read on climate change or even the second. It is important to understand the science and some of the policy and economics of climate change before delving so deeply into our own inner constellations. This book addresses some of the climate science, but only in relation to the inner climate. It addresses the economics, but only insofar as it touches on how we live together. If you do not know much about climate science, or if you need a refresher, you should probably put down the book and go to Wikipedia. Read some articles, and pick it back up. Then again, each of us has our own ways of processing information, so maybe this will be your gateway drug. There are worse ways to be introduced to climate change.

Most books on climate change start by showing how climate science implies worst–case scenarios, like flooded cities, melting icecaps, devastated crops, and water wars. The concerns are real and serious. But many writers frame the issue so as to direct our responses down a narrow channel. More often than not, they signal panic, throwing us headlong into the rapids. Reading such books can be like struggling for air. But just when the average reader seems likely to go under, the authors hold out a branch. We can save ourselves by implementing their program. It is an old formula that can be very powerful. But it involves a momentum that is not conducive to self-reflection and impartial reasoning. The momentum can carry us too far, too fast, as we lose our psychological moorings. The scare tactics create a momentum that is not conducive to strategic planning. And it often ends in the wreckage of our vessels. Quite a number of environmental activists have quit the game believing we are sure to lose. But quite frankly, after two decades of activism, community organizing, and social entrepreneurship, it is an approach that has come to bore me, like a good metaphor that has become a cliché.

Perhaps someone will do a study that finds the environmental activists who last are the ones who finish reading the books they start, while those who burn-out tend to quit half-way. For the dire warnings tend to come first, while the solutions show up toward the end. This book is for those who pick up information wherever they go and constantly question their relationship to what they learn. In the end, this may prove a better way to go. It may make for a nice compliment. And it may just be better suited for a certain kind of reader. Perhaps you are this sort of person. Or maybe this book will show you something new. Good books do tend to take us on journeys somewhere new. But perhaps it will just keep you from becoming desensitized to the highly effective scare tactics a little longer. The Inner Climate was produced in the spirit of embracing diversity, of creating space for more points of view, of allowing us to flourish while we do our best to carry on with work that must be done. This book will be a success if it makes you think hard about your response to climate change and provides a vantage point from which that thinking can be sustained. The interviews have been edited to improve readability. Contractions have been removed, incomplete sentences completed, and unnecessary words removed. While major changes have been approved by the interviewees, final copy-edits were not, and all stylistic decisions concerning punctuation were my own. Readers are encouraged to start with the dialogues that seem most interesting. I hope you will learn as much in the reading as I did in the interviewing.

Theo Horesh,
Boulder, Colorado

SECTION I

MIKE HULME
MICHAEL SHELLENBERGER
PETER SENGE
FRANCES MOORE LAPPE

MIKE HULME

Mike Hulme has authored, co-authored, and edited six books. He has been studying climate change and our responses to it for over three decades, through both the physical sciences and the humanities. He was a lead author for the 2007 IPCC report. And he is the Founding Director for the Tyndall Centre for Climate Change Research, which was started in 2000.

He is perhaps best known for his book, *Why We Disagree About Climate Change*, where he challenges the idea that there is a consensus on how we should respond to climate change. He argues instead that since climate change touches on so many aspects of our lives, our answers as to how we might grapple with it tend to diverge. There are a multitude of different ways we might respond to climate threats. Narrowing the range of acceptable responses tends to generate resistance and results in missed opportunities. Since climate change draws so much into question, it is an opportunity for us to re-examine how we live together.

It is through such an analysis that Hulme takes climate change out of the hard sciences and into the humanities. It is a shift that has been mirrored in his own career, where his work as a geographer has allowed him to span these two often divergent spheres.

THEO HORESH: *You have written, "Perhaps we have out-maneuvered ourselves. We have allowed climate change to accrete to itself more and more individual problems in our world - unsustainable energy, endemic poverty, climatic hazards, food security, structural adjustment, hyper-consumption, tropical deforestation, biodiversity loss - and woven them together using the meta-narrative of climate change. We have created a political log-jam of gigantic proportions, one that is not only insoluble, but one that is perhaps beyond our comprehension." So, how might we break through this log jam and begin a more productive conversation about climate change?*

MIKE HULME: The way I approach it is to redefine the problem. Climate has evolved as an issue, over 20 or 30 years, into a puzzle, a problem, which seems to touch on so many different dimensions of humans, living in a complex world, that if climate change is going to be solved, it almost implies all of our other problems have somehow got to be solved. And that just does not stack up to me as a very likely scenario, that somehow we can find the solutions to all of those problems that you mentioned. I sometimes use the metaphor of a Rubik's cube, which to me is a horrendously difficult puzzle with so many different permutations that I could never solve it. I think the way we frame climate change has reached the stage that it really has those attributes of the Rubik's cube.

What I suggest is that we rethink what is the problem of climate change. And it is, of course, a multitude of much more well-proscribed, limited, and tractable elements. To take one or two of those, there is a problem that we have around contracting areas of tropical forests, with all of those diversities of life and medicinal properties and other ecological functions that tropical forests offer to us. We have a problem around the 1.5 million people, mostly children, who die prematurely each year because of indoor smoke pollution, because their domestic cookers are not electrified. Or we have a problem, as indeed we recognized back in the 1980s, around this category of gases called halocarbons, the CFCs and the HCFCs, which deplete stratospheric ozone. With that one we created a mechanism, a procedure that has gradually eliminated those ozone depleting gases - but not entirely yet.

Those are just three examples of much more bounded concerns, problems which can be tackled and approached on their own terms. And I argue that they could be largely reduced in magnitude, if not eventually solved. And one of the benefits of doing all those things - constraining the rates of tropical deforestation, reducing black carbon from indoor cookers, and eradicating HCFCs – the co-benefit, the side benefit, of those things, would be to reduce the human influence on the climate system. It is doing two things. First, it is breaking climate change down into smaller, more tractable problems. Second, it is turning around the justification for policy innovation. It is not trying to solve or to stop climate change from happening. It is trying to secure other public goods or to eliminate public bads. And the co-benefit of doing that is to reduce human influence on the climate system.

It sounds like what you are saying, to stick with the log-jam metaphor, is that we need to pull one log out at a time, and that is what is going to break this impasse.

Actually by removing one log at a time, there is a benefit, a public benefit, of removing the log, that is over and above the ultimate goal, which might be to remove the jam completely. But before you can get to that ultimate goal, you actually have a number of side benefits along the way, which can act as a motive and justification for smaller interventions around more manageable problems.

So, your book raises the obvious question, why do we disagree about climate change?

That is the title, and what I do is lay out six or seven different dimensions of people's world views, beliefs, ideologies, values, which lead them to see the world in different ways, to interpret evidence in different ways, and which leads them to act on that interpreted evidence in different ways. For example, we disagree about climate change because we attach different weights to unborn generations. Some people will attach much higher weight, morally and ethically, to unborn people than others. Others actually may attach much greater weight to the disadvantaged and suffering who live in today's world than they might to the unborn, hypothetically poor and disadvantaged, of future generations. Now that is a different ethical judgment, a different moral judgment, which no amount of scientific evidence is going to reconcile. It emerges from people's different belief systems, different ethical orientations, and that is one of many reasons why we will disagree about what climate change signifies for the world today and how we act in response to it.

What is it about climate change that opens up all of these different fronts where we might disagree?

That is a really interesting question: what is it about this phenomenon, because it is rather different in character from a number of other categories of environmental problems. And that was certainly how climate change was originally framed back in the 1980s, as an environmental problem. But it takes on a rather different character than, for example, the dangers of asbestos in buildings or the depletion of ozone in the stratosphere, which I have mentioned already. It takes on a different character than say mercury poisoning in the food chain. So, what is it about climate change that makes it so much more multi-dimensional and deeply embedded in all of our human systems? I think there are two parts to that. One is that the drivers of human-induced climate change are largely energy use and our use of land. They effect virtually every part of economic activity in a way that asbestos does not. Chlorofluorocarbons and the ozone problem were actually quite a circumscribed problem. Climate change touches on every aspect of economic activity. The other thing about climate change, that gives it a much more intricate and intractable character, is that it deals with something that all human beings feel very deeply about, and actually feel very attached to, and something that all human beings feel that they have got personal knowledge of, which is the weather. In the end, the risks of climate change are first and foremost hazards related to weather that occurs in particular places, at particular times, and affects particular people. They are hazards which are culturally mediated and socially differentiated.

Human cultures have evolved over generations to have very intimate relationships with weather. That, again, puts climate change in a different category than one of those previously mentioned issues. It is really only in the last two generations we even knew there was a layer of ozone in the stratosphere. Asbestos was a manufactured building product that was not known about 150 years ago. The weather predates humans, and our whole physical and cultural evolution has developed very specific relationships to it, both in terms of our practices and in terms of our stories and our rituals, our psychologies, our emotions. We feel very attached to our weather.

There are numerous issues with which we might concern ourselves: world hunger, disease, loose-nukes, over-population, deforestation, loss of species, depletion of the oceans, economic inequality, just to name a few. Climate change has taken center-stage for many us, largely because it is so deeply tied to many of these other issues. How do you situate climate change among these many worries, and what is most important about it to you?

My thinking has changed over my years of studying climate and society, which is a career going back 30 years or more. And again, in the early stages of my work around climate change in the 1990s, I probably would have seen it as this over-arching problem that really did sit at the top of the pile, as far as public policy concerns went. And I think what I have realized or come to understand is that it is just too big a problem to take on at one go.

We can choose our problems - problems are not somehow mandated to us by some inner-voice, speaking inside our head. It is people in societies who choose the problems they wish to tackle. Until the eighteenth-century, we chose not to tackle the problem of slavery. Until the mid-twentieth century, in North America, the problem of civil rights was not chosen to be campaigned against. Problems are chosen, or politically negotiated one might say, and I think that by choosing to tackle climate change 25 years ago we actually chose a problem that was too big for us. It does tie-in, it does connect with, all these other things you mentioned, but we are not going to solve climate change by putting it at the center of our universe of campaigning or policy innovations. We need to take a couple of steps back, rethink what some of the constituent elements of this problem are, and tackle those. And in the process of taking those issues on, we will in the end reduce the scale of human influence on the climate system.

What does "solving" the question of climate change mean to you?

I do not think I would even recognize what a solution to climate change is now. What I would do is look for solutions or interventions that reduce the size of human influence on the climate system, and the solutions would emerge from tackling some of the other constituent elements. For example, one of the ways I break climate change down is to focus on weather hazards and the danger that weather extremes present to humans and the things they care about. Those weather hazards - flood, drought, heat, high tides, and so on - exist and have existed for all of human history. Human activities, most likely, are beginning to change some of those weather hazards. But the point remains: weather is dangerous, and so one element of tackling climate change is the need to focus on minimizing the dangers of weather, for particular people in particular places. That is not solving climate change, that is not stopping climate change. But what it is doing is minimizing the risks and dangers that weather hazards present to humans and the things they care about. The precise extent to which those weather hazards are influenced or caused by human activities is actually a secondary question. The primary question is how we reduce those hazards and risks.

Looking at this question of the looming nature of climate change raises the corresponding question of the end of civilization, the end of life on earth. One of the things that seems most absent from discussions on climate change is a serious exploration of how we grapple with it psychologically. What happens when we begin to confront the potential end of the biosphere, all the environmental systems upon which we depend for our survival? How does this affect our ability to have a rational discussion about the impacts of climate change?

In the way you have posed that question, I think you have jumped two or three steps ahead of where we are and what knowledge we have about the future. I know that, for some, climate change triggers, immediately, these pictures of a global catastrophe, the ecosystem collapse, of human survival, even civilizational survival. I cannot automatically make those connections. It seems to me there are least a couple of phases of reasoning which you would have to go through before you can get from what we know scientifically about what is happening to the climate, and our influence on it, to those consequences. So, I do not think in terms of biosphere eradication or destruction of civilization or the elimination of the human species.

What I think about is this intricate relationship that humans have with their non-human surroundings, with their physical surroundings, and how that relationship has always been. Since the very earliest years when our ancestors began for the first time to intervene deliberately with primitive technologies to alter their physical environment in a way that the other primates were not able to do - maybe 30,000 to 50,000 years ago - humans have been changing and altering their physical surroundings. And the scale of that intervention has grown and grown. The sorts of systems that we have now left our fingerprints on have expanded, from the local clearing in the forest to the rivers and the lakes and the oceans and now, with climate change, to the skies and beyond.

But this is a story of continuity rather than a story of disjunctures. And all the time, as humans have extended their reach and influence, they have changed and transformed the physical world and by their social institutions created uneven access to these transformed resources. So, the large parts of the earth's surface we are now living in are at best what one might call hybrid or novel ecosystems. There is nothing preternaturally natural about them any longer. They are co-produced, between physical processes and human activity.

The way I look at the future is not to fence these psychological shocks that you referred to, thinking that at some point humans will become extinct because of what we are doing. It is simply to grapple with this continually evolving, co-evolving story of how humans interact with their environment and how power over those resources is distributed. And that presents challenges, it presents hazards to humanity. And it presents opportunities and necessities of innovation and new systems of organization.

By the early nineties, numerous narratives and philosophies had arisen to explain our ecological crisis. Psychologists, sociologists, philosophers and theologians were all beginning to make substantial contributions to making sense of the relationship between humans and nature. I am thinking here of Deep Ecology, the New Cosmology, Social Ecology, Eco-psychology, Wilderness Therapy, and Eco-feminism, to name a few schools of thought. How has this shift to climate change impacted these ways of thinking, and what sorts of new world-views has the shift to climate change generated?

Because climate change has become such a pervasive idea in contemporary culture, over the last 25 years, you can go to pretty much any country in the world and, within a few hours of arriving there, find yourself in conversation with people about climate change and what it means to them. So, this is a very pervasive idea that has reached across and entered into pretty much all cultures and societies. It does not always mean the same thing in these different cultures, but it is an idea that one can have a conversation about. And so I think one of the things climate change has done is to make it much more visible and inescapable for people to recognize this close relationship, this interdependency, between humans and their non-human environment.

Those philosophies that you listed would have been seen as rather niche and obscure, maybe 20 or 30 years ago. The idea of climate change has given them more visibility - not necessarily more acceptance, but certainly more visibility. Climate change offers a new vocabulary to conduct those philosophical and spiritual conversations. I think the other thing that climate change has done is to draw attention to this long human trajectory of ever increasing presence in the non-human world and now to changing the functioning of the planetary climate. The human fingerprint, human reach, extends even to the skies, and the circulation of the atmosphere, which is, of course, an ultimate planetary intervention - I think it helps us to understand that story.

And then going back to what we were talking about before about the psychology of fear, there are at least two ways you can respond. One is to shrink back and to say, "This is not good, this is wrong; we do not want to have this scale of intervention." The other response is to recognize that story and say, "This is the nature of the human." And in a way, this is where the idea of the Anthropocene, the coinage that Paul Crutzen and others promoted about 10 years ago, offers a particularly vivid way of thinking about this. The Anthropocene is an era in which humans are planetary agents of change and transformation, but some humans very much more than others. For me, certainly, there is no getting away from that, and to pretend that somehow we can de-invent the Anthropocene, or we can back out of it somehow, is, to me, actually quite illusory thinking. We cannot go backwards in that sense; we can only go forwards, and recognize that this is, indeed, the nature of the human; that through technological innovation, through population expansion, we have got an aspiration for more material through-put. We are this being; this is who we are. And we have to face up to that reality, rather than shrink back from it.

Throughout your writings, you have emphasized how thinking deeply about climate change might inspire spiritual and cultural renewal. You have also laid out a wide spectrum of axes upon which we might disagree about climate change. What is it about this minefield of differences that makes climate change such a fertile field for growth and renewal?

Climate change presents the scale of the human presence on the planet and the future of this ongoing story of co-evolution in very vivid terms. I do not think it is something that we can just pretend does not exist. We may have different responses to it, and that is why we disagree about what our responses are, but climate change demands we face up to that reality. And facing up to that reality does ask quite fundamental questions about the future, about what we see as the purpose of the human story or the purpose of humans themselves. And it asks us, and challenges us, to think where we find meaning for our own lives within this unfolding story. Those are the questions humans have always asked, and that is what has prompted this very powerful sense of the transcendent and the spiritual in the emergence of great world religions. Climate change actually makes those types of questions very much more pertinent today than perhaps they may have done in the middle of the twentieth-century or perhaps even going further back to the beginning of the twentieth-century, before the beginnings of disillusionment about the narrative of progress and onward advance of greater human triumph.

Now climate change asks these questions in fresh terms. These are questions that resonate down through human evolution, but we now are presented with them in a fresh way, with new vocabularies and new questions about what is the actual purpose of humanity. What is my contribution to that purpose, is it about survival, is it about securing greater welfare and well-being, is it about maintaining somehow the particular biodiversity of life on the planet? Because all those things are now thrown up into the air, as things that are not simply going to happen through nature but are going to be decided upon by human actions enabled or constrained by relationships of power. And that then forces us to ask these types of transcendent questions, which themselves do not find answers simply by commissioning new scientific research. They have to come out of other forms of reasoning and reflection and even creativity.

Listening to you speak and reading your writings, I get the sense that you have not just done a lot of research as a scientist, and in the humanities. I get the sense, rather, that you have meditated really deeply on the problems of climate change, that you have let this infiltrate your life and felt into it deeply. Part of what I am getting at is the uniqueness of your viewpoint - it feels like you are touching on issues that matter, that are not being touched on by many others. Could you say a little bit about the development of your own thinking on climate change.

In my undergraduate training in geography, 30 or more years ago, I was, obviously, introduced to many of the great geographical themes of the planet and human habitation of the planet and climate and its variability over the ages. And its current variability, of course, was part of that induction. But I saw climate change as quite a discrete phenomenon, as we have touched on already. Then, as I entered my professional career as a scientist, I studied climate change for 15 years or so very much through the physical sciences, particularly working on observational data - what the data could tell us about trends of the changing climate; and then later, with collaborations with modeling groups - how possible it is to simulate with models these variations and changes in climate; then later than that, to try to apply those observations in the models to the field of scenario-development to inform policy and decision makers. But all along I saw climate change as a rather discrete phenomenon. Since I set up the Tyndall Center back in 2000, and deliberately sought to expand the range of disciplinary perspectives that we worked with within the Tyndall Centre, and then even more so, when I stepped down seven years ago and explored much more what some of the humanities and interpretive social scientists were thinking about climate change, I began to find ways of bringing together my own personal commitment to knowledge and being.

As a religious believer, a mainstream Protestant Christian, I began to try to see how all these things actually fit together. Climate change is not something that can be put in a box, whether a scientific box or economic or even a policy box. This is really asking some pretty powerful questions, which certainly I recognize from my own Christian religious tradition. And that is part of where I am at now - trying to find ways in which these things somehow fit together to make a much bigger story, which allows us to bring in religious thought and ethical reasoning and historical perspective and creative culture, alongside economics, policy, and science. But if we limit it to economics, policy, and science, then I think we are actually missing the larger account of humanity now in the twenty-first century. We need to spread our canvas much more widely.

We now have environmentalists, architects, entrepreneurs, paleo-climatologists, and restoration ecologists, not to mention all of the academics, lawyers, engineers, and venture capitalists, building good careers on being Green. Noticeably absent amongst these roles are a comparable number of representatives of the social sciences and humanities. Can you speak a bit about the challenges of bringing environmental awareness to the humanities and the humanities to the environmental movement?

Actually, in the creative professions, we have seen over the last decade or so a much greater engagement and openness for artists and poets and novelists and philosophers to work with this phenomenon of climate change and to do creative things with it. The art exhibitions and novels that are now inspired by and grapple with some of these themes that we have been talking about are much more present today than they were just 10 years ago. It is very positive and very encouraging that people recognize now that you cannot make sense of climate change simply by dealing with numbers and models and measurements. Actually, to make sense of climate change, you have to bring in those professions and traditions that work with the interpretive, the imaginative, finding ways of expressing what things might mean to people, which is what the humanities disciplines and professions are skilled at doing. Within the academy, I am increasingly keen to work with historians and philosophers and literary critics and religious scholars to enrich our understanding of what climate change means.

Working with so many disciplinary perspectives complicates the story. It begins to show that there are many different stories at work around climate change. It is too limited of an ambition to simply stop climate change at 2 degrees of warming above pre-industrial levels. By working with these larger humanities disciplines, we begin to find other ways of telling the story of climate change and how we might envisage or imagine the future.

It seems like there is a paradox in thinking about climate change, that on the one hand it is such a big issue, spatially and temporally, that we need math; we need numbers to sort through what is happening. We need to measure chemical compositions of the atmosphere and what not, and we need to be doing projections into the distant future about what the changes we are seeing now are going to bring, both atmospherically and economically. And at same time it is also so big that the vast majority of people cannot really grasp what those numbers mean, and so we have to bring in this more imaginative perspective. What concerns me about getting too imaginative with climate change, and getting too local in our solutions, is that we will lose our grasp on the big picture and get sidelined by tangential issues. I am wondering if you can speak to this question of how we balance the statistical measurement of climate change with the imaginative embrace of things that are meaningful to us as individuals and small groups.

That question draws attention to one of the central dilemmas at work here. You talk about the global analysis and enumeration of the planet and temperatures and sea levels and concentrations of different chemical species in the atmosphere, offering these sort of global indices and indicators, suggesting that they may be heading in directions we do not particularly want them to. But then what we do not have at those global scales, at this perspective, is any ability or capacity for decision-making or political action. So, one of the reactions to that frustration and deficit is that people will turn to scales of decision making and political actions where there are tried and trusted institutions and procedures for doing politics.

There are at least two ways you could react to this political deficit. One is to say, "Well, therefore we will need new global political institutions, that have a mandate and a scale of perspective and activities that is planetary and commensurate with these sorts of scientific indicators that we have mentioned." The other response is to say, "That vision is idealism run rampant: it is sheer illusion to somehow think that, given the deep differences and divides that exist between nations, and are expressed through political institutions at the present through the nation-state, any democratically mandated political institution could possibly act on a global scale." So, I think there are different ways to react to that dilemma, to that political deficit. It actually throws into sharp relief the reasons why I have retreated from those ambitions to solve climate change through global multilateralism, one-size-fits-all solutions, the Global Deal. I just do not see that as politically achievable. So, I find myself preferring to design the problem to match the political capabilities that we have for solving problems. It is pragmatism over idealism; it is "not to let the best be the enemy of the good."

MICHAEL SHELLENBERGER

Michael Shellenberger is an environmental strategist and a leading global thinker on energy, climate, and human development. He is currently the President of the Breakthrough Institute, which he co-founded with Ted Nordhaus in 2003, and which advocates government involvement in clean energy innovations. In 2007, Shellenberger and Nordhaus co-authored *Breakthrough: From the Death of Environmentalism to the Politics of Possibility*, which was described by Wired magazine as "the best thing to happen to environmentalism since Rachel Carson's Silent Spring." The book inspired a national debate over the basic values and strategy of the environmental movement and in 2009 was awarded a Green Book Award, previous recipients of which include E.O. Wilson and James Hansen.

Shellenberger persistently challenges environmentalists to embrace economic development. He argues economic development is necessary for alleviating global poverty and is integral to human development. By embracing it, the environmental movement can forge stronger progressive coalitions in the developed world and garner support from the Global South. Economic development also spurs an interest in higher values, like the love of nature and a willingness to sacrifice for future generations. And it will be needed to bring about the clean energy innovations. The New Republic praised this work as "among the most complete efforts to provide a fresh answer"how liberals might modernize their thinking. But Shellenberger and Nordhaus have not just sparked debate. In 2002, they founded the Apollo Alliance, now the Blue-Green Coalition, which advocated that the government invest $300 billion in clean energy technologies. The goal quickly moved from the margins of environmental thought to become a major policy initiative which was adopted by the Obama administration. And in 2008, Shellenberger and Nordhaus were named by Time magazine as two of its 32 Heroes of the Environment.

THEO HORESH: *You have been a tireless critic of traditional environmentalism and yet an increasingly pivotal leader in the environmental movement. As an old saying goes, "a true friend is someone who will stab you in the front." And there are few things as uncomfortable as telling a friend how their behaviors have become self-defeating. This seems to be what you are doing with the environmental movement. Can you talk a bit about your relationship to the movement and some of the changes you hope to inspire in it.*

MICHAEL SHELLENBERGER: There has always been a very negative view within the environmental movement, that dates back to before Rachel Carson, of modern life, modern capitalism, prosperity, which is viewed as something that is really terrible and needs to be reversed. You have got a strong strain of the environmental movement that has long wanted to go back in time to when things were simpler, when technologies were simpler, when societies were less developed. We have been attempting to speak more critically of that strain of the environmental movement, in part out of a commitment to human development and human freedom and human flourishing.

We find that discourse is strongest in the wealthy parts of the world. And what we have been trying to do over the last 10 years is draw attention to the fact that most of the world does not consume enough energy, does not consume enough resources, and that for the rest of the world to live modern, prosperous, fulfilling lives we are going to need to use a lot of technology. This posture really embraces our affluence and our development, while also understanding that if we are going to have a world of 9 billion people, all of whom are consuming at least European levels of resources, we need to do a lot more technological development so we can protect our ecological heritage.

Your perspective captures so much more of human needs and aspirations. But it seems that the traditional environmental critique of West-

ern, developed, capitalist society is that this society leaves so much out. It leaves out a connection to the non-human life-world and to the parts of ourselves, maybe unconscious parts, that are more wild, beyond our intentional sense of self-control. Do you see this embrace of the wild, of that which we cannot control, still playing some part in your vision of the environmental movement?

You are exactly right. One of the things you hear a lot is that we have lost touch with nature. But I think the opposite is the case. My mom, who grew up in rural Indiana and was born in 1939, the youngest in a family living off a farm, felt too close to nature. Polio was still rampant. They were engaged in all sorts of back-breaking labor. The nature they experienced was not that of walking through redwood parks or visiting pristine shorelines. The kind of experiences that we now have with nature, as wealthy moderns, are so much more positive than the experiences people were having with nature when they were much poorer. So, it is a funny complaint.

In some ways the automobile, the interstate highway system, all of our wealth that came strongly after World War II, all of that opened up possibilities for us to experience nature in a much more positive way. There has been some confusion and some forgetting about just how hard life was, and how hard nature is, when you are poor. So, part of what environmentalism comes out of is a period of astonishing amounts of wealth. In this time, coming out of the fifties and sixties, the idea that capitalism immiserates the poor, which is an idea that dates back to Marx, gets disproven for a lot of folks on the left, in that American workers, even the poorest, become wealthy consumers.

What you saw with environmentalism in that period was a turn to argue, "We might be materially richer, but we are spiritually poorer," but I don't buy it for a minute. We have actually become much richer spiritually. You see a lot more diversity in terms of spiritual communions with nature, a lot more ability to actually experience nature in all of its different forms, when you have access to high levels of energy and a modern life.

Your writings do not seem to fit into any kind of traditional category. Sometimes your book, Break Through, reads like a postmodern critique of modern conventional environmentalism. You deconstruct the idea of nature, the notion of environmental justice, the importance of the pollution paradigm. At times you sound like what to many people would be a conservative, criticizing environmentalists for their lack of respect for ordinary people and basic creature comforts. But what you seem to have laid out in Break Through is the foundation for a comprehensive, progressive movement that includes far more goods than progressive movements have tended to include in the past several decades, perhaps far more than we have ever included in progressive movements in America. What was it you hoped to accomplish with Break Through, and several years on down the line, how much of your vision is being actualized?

We sometimes find ourselves in a lot of agreement with liberals of the era before say, 1968. There is a mythology that before 1968 nobody cared about the environment. But the more you read of that history, in fact, you find people were constantly concerned about the environment. A number of clean air and clean water acts were passed in the fifties and sixties, in fact, before then. One of the interesting cases we have been looking at is the Tennessee Valley Authority, which we usually think of as a program to provide electricity for poor people. What we forget is that the TVA was really an environmental restoration program. They were experiencing severe deforestation in the Tennessee Valley region, because they were reliant on wood for fuel, and they did not have modern fertilizers and modern agricultural technologies. So, what the TVA did was it built dams, which became viewed as a negative for the environment. But the dams provided electricity, which provided fertilizer, and they also did tree planting, reforestation. The combination of those things, starting back in the thirties and forties, actually improved the environmental quality of the Tennessee Valley region dramatically. So, we wanted to get environmentalists back in touch with a tradition of American liberalism, which says government has a role in modernizing and even intensifying energy and agricultural sectors.

Instead of dedicating more of nature, the non-human world, to human needs, you actually want to reduce the human footprint. That is where we basically agree with mainline environmentalists. What we disagree with is that the best way to do that is through organics and renewables alone, because such an approach actually requires using more land for agriculture and energy. We want to intensify energy and agricultural production so we can set aside more of the earth for non-humans. And so one of the biggest changes we have had since Breakthrough, that is starting to settle in, is the idea that you want to use technology to intensify agricultural and energy production so you can leave more of the earth for non-humans; that is what you want to use technology for. What you do not want to do is spread agriculture all over the planet, because agriculture is arguably the most environmentally disruptive thing humans do.

So, by intensifying some of the core aspects of economic development, particularly energy production and agricultural yields, we can reduce our footprint on the earth?

That is exactly right. In fact, it is what has been happening already. Since the early sixties, the amount of land that we dedicate to agriculture has actually stayed the same worldwide. And yet, over that time food production and yields have gone way up. That is an incredible achievement, and we should continue to do that so we can reduce how much land we are dedicating to agriculture in the future, and we can reduce how much land we dedicate to energy in the future. And so, we have been really interested in that process of modernization, which gets called "intensification" rather than "extensification."

There is a deeper concern many feel, that there might be something you are missing, that this whole economic developmental process can go awry, and that we are going to eat up all the resources on the planet and overfish the oceans and destroy the world's remaining wild places. How does this concern fit into your vision?

We share the concern that we are going to use up too much of the earth or that we are going to degrade it and are degrading it in ways that are really negative. So, I think you have to pull some things apart there. The biggest human impact on the planet is agriculture. About 11 or 12 percent of the earth's land mass is dedicated to agriculture; less than 1 percent is dedicated to energy production. So, the first thing you want to do is to shrink that footprint, and the way you want to do that is through agricultural technology. The other thing is you want the human population to decline or to not continue to grow so exponentially. And that is very likely to peak at 9 billion, maybe 10 billion, and then go down. Declining fertility is strongly correlated with increasing human development, urbanization, education of women, empowerment of women, and even television, as it turns out - so that is a positive trend. You want more people living in cities and fewer people living in the countryside. That is also one of the dynamics of modernization.

The thing you want to pay attention to are the trends that are positive, that you want to increase. In other words, how do you use your strengths to overcome your weaknesses? The way this sometimes gets painted by environmentalists is that humans are a sort of cancer on the earth. There is nothing controlling our impact; it is just out of control. And I think this is a false picture. There is a whole set of ways in which our impact is actually diminishing. One little fascinating tidbit is that by far the vast majority of deforestation occurred before 1800. The reason is obvious: we greatly reduced how much wood we use for energy. We tend to think of coal as this terrible thing for the environment, but coal allowed us to stop using wood. And when you look at these energy transitions, we go from burning wood, to burning coal, to burning natural gas, to using nuclear or renewables - each of which has a lot less pollution, a lot less environmental impact. That is progress, and the goal is to accelerate those transitions, make those transitions happen faster and in ways where we minimize the unintended consequences.

Break Through seems to be a bit of a triple entendre. In one sense you are breaking through old ideas that have run their course. They no longer can address the issues at hand, particularly climate change. In another sense, you are talking about breakthrough technologies, and there is a real belief in supporting and fostering breakthrough technologies. Yet in another sense, there is this whole thrust of the human spirit; we are breaking through to higher developmental aspirations. And this whole energy, the associations tied to breakthrough, seems to grate against a more traditional environmental paradigm. Can you speak a bit about the, no pun intended, thrust behind Breakthrough?

You really got it right - it is a book that argues for dealing with environmental problems by going forward rather than going back. And you are right about the double meaning. One of the things that you learn about technological breakthroughs is that they take a long time; it is one of their paradoxes. They are not made to order. They often take decades to achieve, and that means you have got to start right away. The underlying assumption of *Break Through*, that we have spent a lot more time working with at the Breakthrough Institute, is the idea that we just do not have all the low-carbon technologies we need in order to replace fossil fuels. We have technologies, they are just not cheap and scalable. And some of them like nuclear are opposed by the same people who are most concerned about taking action on climate change. So, you have a set of technological problems that have to be solved.

One of the big policy ideas that comes out of *Break Through*, and that we still strongly believe in, is that the government has a really important role to play in advancing the development of those breakthrough energy and agricultural technologies.

Can you talk a bit more about the role of government investing in new technologies, but also about the role for markets in what you are advocating. It seems like what you are getting at is a sort of vigorous government involvement, and a vigorous use of markets, which really breaks through not just the environmental movement's, but also the progressive movement's, regulatory perspective.

The biggest difference between us and most environmentalists seems to be our view that the government is the most important actor when it comes to technological progress. It is often private companies that are doing the work, but when society decides it wants to solve a particular problem - whether it is polio, or national security against the Soviets, or getting more natural gas out of the ground, or creating a next generation solar and nuclear technology – it is the government that society turns to in order to push those technologies. And the government will often hire private firms to do the research and development, the demonstration, and then the deployment of those technologies.

The story most environmentalists have had in their head for 40 years is that technology development occurs in response to regulations. And there is some evidence for that, but it just pales in comparison to direct government support for innovation. In the worst kind of manifestations of this you find a lot of green groups that criticize governments doing technology innovation. They call it corporate welfare. They will say things like, why are we funding natural gas developers to try to get natural gas out of shale? Or why are we funding the nuclean industry to develop next-generation nuclear technologies? It is a bit of a double standard, because green groups do tend to defend subsidies for solar and wind.

You did not get the personal computer by having a cap and trade system on typewriters. You did not get the Internet by having a tax on fax machines. You got these things by just directly developing them, by directly demonstrating them. You had a lot of trial and error, certainly, a lot of waste, sometimes even corruption. But after several decades of it you got all of the major technologies in the iPhone, which does not really happen without 40 years of Defense Department investment in technology. Every major energy technology comes from long-term government investments in innovation. And we are trying to remind the environmental community and liberals that this has been the fundamental role of the federal government for almost 200 years. We need to reaffirm that role and put it in the context of improving environmental quality.

And you say the fundamental role of government?

I do - I think it is fundamental. Private companies are just really limited in what they can do to innovate, especially now that there are all sorts of pressure on CEOs to deliver higher earnings right away. There is a lot of short term thinking. Real breakthrough innovations take decades. And often guys like Steve Jobs are having to do things that their colleagues see as a waste of money. Even Steve Jobs was limited in what he could do. He needed to have Xerox Park, which was the famous Xerox R&D Laboratory, to develop those technologies. And Xerox Park needed Defense Department contracts in order to develop all those technologies. So, let's remember that these technologies come from long-term government support, and there is really no substitute for it. When we put our minds to dealing with the various environmental issues we are concerned about - climate change, habitat destruction from agriculture, fresh water scarcity - when we put our minds to solving those problems through technology, then we discover a really important role for government that does not have anything to do with regulation, does not really have anything to do with markets for that matter either.

Let us say we get vigorous government investment in technological development: what role do you see markets playing in developing the clean-energy economy?

Markets are ultimately the thing that proves whether or not your technology is viable. We see it shaking out most dramatically in the case of solar, but also in nuclear. You have got some really exciting solar and nuclear technologies, but they are not able to compete against natural gas. And so for those technologies to be able to compete against natural gas and coal, they are going to have to become a lot better and a lot cheaper. So, you have to have the discipline provided by the market.

But it is funny, because you always hear people say, "Nuclear or solar just cannot compete against natural gas." And that is true, but natural gas could not compete against coal until a few years ago. Technologies change and they develop. Who would have thought that natural gas would be cheaper than coal? People would lecture us and tell us, "Your vision of making clean-energy cheap is never going to work, because nothing will ever be cheaper than coal." Well, natural gas became cheaper than coal, and low and behold, U.S. emissions started going down, beginning in 2007. Emissions in the United States have been going down faster than in any other country in the world, and it is almost entirely due to replacing coal with natural gas.

Let's talk about natural gas. Most major environmental organizations have been opposed to hydraulic fracking, which has allowed us to get this cheap and abundant natural gas, which has allowed for CO_2 emissions to go down significantly in the United States. And at the same time there are massive concerns with methane leakage, which many suggest might actually increase total greenhouse gas emissions, since methane is such a powerful greenhouse gas, thereby raising global temperatures significantly - at least in the short term. How do you respond to these criticisms of natural gas? How vigorously and under what circumstances do you want to support and get behind natural gas?

It is important to look at the facts: according to the newest EPA numbers, methane leakage actually declined 10 percent since the early nineties. Over the last 20 years, methane leakage actually went down - even though gas production increased by more than 30 percent. Why is that? Preventing methane leakage is not hard, it is not a huge technological challenge. It is not as hard as say turning nuclear waste into fuel or capturing and storing carbon dioxide underground or making a solar-panel much more efficient. It is a pretty straightforward technological fix. And the other advantage, of course, is that the methane that leaks is the product, it is the natural gas. So, the industry itself has a strong interest in preventing the leakage, and so you see methane leakage go down even as gas production has gone up.

We just put out a report where we looked at all the big environmental consequences of energy production, and we just compared natural gas to coal. We looked at everything: we looked at methane leakage, at water pollution, at the different kinds of air pollution, sulfur dioxide and nitrous oxide. We looked at mercury, at mortality, human deaths per unit of energy. And gas was superior to coal on every measure and often by an order of magnitude.

I saw the report and I agree.

It is a funny thing, because there is no question that gas is an environmental improvement over coal. There is no question that gas is an improvement for human lives over coal. Does that mean that natural gas is without consequence? Of course not. They are fracking in my hometown, under my elementary school. But I come from a conservative town in Colorado, and mostly people there are in favor of fracking. You go an hour up the road to Boulder or to the more liberal parts of the state and people hotly oppose fracking.

So, a lot of what you are seeing is a reaction to fracking from people who really do not want the industrialization of their country homes and the countryside where they live. You see a reaction to it by people that have a vision of moving right from coal to solar panels and wind. And I appreciate both of those concerns, there is merit to them. But I do think the question that we are being asked is, "Do we want to replace coal with gas?" We are not choosing between gas and some other energy source as a replacement of coal. And when you understand that is the choice, it is a pretty straightforward one to make.

You have talked about the declining footprint of our major energy sources. You started with wood, but we could even start with the human animal, moving on to domesticated animals, going on to wood, to coal, to natural gas, to nuclear, with the later sources becoming progressively more productive and leaving a smaller footprint. But we get to nuclear, and the idea that it is a cleaner energy is just head-spinning for a lot of environmentalists, especially the older ones, who worked so hard in the eighties to shut down nuclear plants. It was one of the great successes of the environmental movement, but here we see a number of environmental leaders like yourself coming out in favor of nuclear energy. And there still are some very real concerns with nuclear; they tend to focus on the possibility of meltdowns and the storage of nu-

clear waste. Then we have the environmentalists, who in the eighties, argued that the biggest problem with nuclear was its cost, that are still continuing to bring up this issue. How do you address these concerns?

Nuclear is such an interesting case. The thing most people do not understand is that nuclear-energy is incredibly power dense. In other words, measured by how much energy you get out of a mass of material, uranium is off the charts, something like three or four orders-of-magnitude more energy dense than natural gas. But when you look at the history of energy modernization - from wood to coal, coal to gas, gas to nuclear - you are moving towards fuels that are both more power dense and have less carbon. So, it is a 200 year process of what is called "decarbonization," which is just a big word for describing less carbon per unit of economic growth. That is the broad trajectory.

Nuclear is so interesting, in part, because it challenges the idea that all economic growth, or all energy, is the same - it is not. And there is this very simplistic idea in the early environmental movement that economic growth is inherently degrading of nature. It is inherently degrading of nature if it requires using wood for energy or if it requires the use of coal for energy. But you can have a world of nine-billion people getting the bulk of their energy from nuclear, maybe using fuel-cell cars, which are highly efficient and non-polluting hydrogen cars. And you can have less impact than you have today, certainly less impact than we had when we were relying strictly on coal and wood for our energy.

There are a bunch of issues with nuclear. The biggest one by far is safety. That is the issue that makes people worry about it, that makes people fear it. And then there are the issues of waste and proliferation. On the issue of safety, the first thing to understand is that when you look at the record of various energy technologies, nuclear is the safest after wind. In other words, in terms of best per-unit of energy, the peer-reviewed scientific studies are all very clear that nuclear is the safest, except for wind. The problem with solar, most people do not realize, is that making solar-panels is an incredibly toxic process that produces a lot of toxic emissions. It uses heavy metals, it uses rare earth metals, and it is actually quite destructive.

We have had three major nuclear accidents - arguably two, since nobody was injured or exposed to dangerous levels of radiation from Three Mile Island. You had a bad accident at Chernobyl, and you had a bad accident in Fukushima. But even there, it is shocking to find out how few people actually died at Chernobyl. There is some debate about the radiation, but at the end of the day, the public health community thinks a few thousand people might die prematurely from cancer, while about 50 people died from the accident. Regarding Fukushima, this is a country where 200,000 people were instantly killed, instantly had their lives taken from them by a tsunami - and then they had a meltdown. I think the trauma of the tsunami really affected how people dealt with the meltdown. They did an evacuation, it was a terrible event, it should never happen again. But amazingly, you had a very small number of people who received dangerous levels of radiation exposure.

So, the solution to the safety issue, in our view is that you have got to move to a different kind of coolant. Basically, all of the problems of nuclear plants today actually deal with a loss of coolant, a loss of water as the coolant. And we think there are better coolants that can be used: liquid metals, molten salt. And these technologies are there, but they need a lot of development. They need a lot of innovation. So, again, you are back to needing innovation to make nuclear safe and make it a lot cheaper.

We think the waste issue is going to be resolved. Most countries in the world have a central repository for it. Ours was supposed to be in Yucca Mountain, but it was imposed by the federal government on Nevada, and Nevada did not want it. Americans do not work that way: we like freedom of choice. But that looks like it is going to be resolved. There is new legislation in the Senate, that is bi-partisan, that would allow states to compete for the waste. We then provide a better incentive.

And then the proliferation issue. In some ways that is one of the biggest psychologically, because people seem to think that if you have more nuclear energy, you are going to have more nuclear weapons, and that is just not the case. There are something like over 30 countries in the world that could make a nuclear bomb if they wanted one. Only something like nine have chosen to do that. Seventy-five percent of the world's emissions are created by the 20 largest economies, most of whom have nuclear weapons capability. So, increasing the amount of energy we get from nuclear is not going to require ready knowledge of how to make the bomb. And frankly, countries do not decide to get the bomb because they happen to have a nuclear power plant. When they want the bomb, they want it for national security reasons. And almost always, such as we are seeing in the case of Iran, they do not hide it very well. But they develop their nuclear weapons through research reactors. In Iran's case, they are calling it a "medical research reactor." They do not even try to hide it as a nuclear power plant. So, I think the proliferation issue is a non-issue. The waste issue is the one that is being worked out, but I do not think it is as big of a deal as most people think. And the safety issue is one that is just going to require innovations if we are going to really move to safer and cheaper designs.

I often wonder if the whole idea of saving the earth, that has so characterized the environmental movement, may have actually come out of a concern with saving the earth from nuclear devastation, and that we have just forgotten the origin of our fears, because we are living in a much more peaceful world now. But we have two countries now – Russia and the U.S. – that could basically destroy the biosphere if they got in a nuclear war together. This of course, seems quite unlikely now, although we never know what the next 50 to 100 years might bring. It is easy to see a situation like the build up to World War I, or the build up to World War II, where the whole game changes, and these 30 countries that have nuclear capability, which most of them are now choosing not to use, could rapidly turn course, though.

Does having an abundance of nuclear scientists and open uranium mining operations and well-developed nuclear technologies make it easier to convert from military to civilian uses? And if a situation like the build up to a major world-war were to occur for a decade or two, and all of these nuclear-capable countries and more were to develop the nuclear weapons technologies to scale, how would we draw hard lines between civilian and military nuclear-scientists, nuclear-technologies, and uranium operations? And what kind of international regulations might we use so that we could prevent such a build-up from happening in the first place?

I think you are right, and I also grew up before the end of the Cold War. I remember watching *The Day After* on television when I was like 12 and it scared the bejesus out of me. So, of course, nuclear war is the scariest thing. An asteroid hitting the earth and nuclear war are human-extinction events, and we are really right to be concerned about them. A question with nuclear technology arises, though: do you think we are going to put the genie back in the bottle? The early anti-nuclear advocates wanted to do that but pretty quickly they figured out it was not possible. Richard Rhodes, who won the Pulitzer Prize for *The Making of the Atomic Bomb*, was very clear about it at the beginning of his book. So, when you understand that we are not going to be able to get rid of that knowledge, we are not going to be able to collectively forget how to synthesize nuclear reactions and make bombs, we are in a really different place. It means that, at least for the foreseeable future, we are going to have nuclear weapons. So, when you ask Richard Rhodes or you ask some of these nuclear anti-proliferation experts, what we should be doing, they talk about reducing the size of the arsenals, taking more safety measures, turning the missiles away, getting rid of the minutemen, getting rid of the kind of nuclear weapons you could launch from jeeps - taking a set of safeguards to really reduce the risk. None of that is a guarantor that we are never going to have a nuclear war. But it is striking that the closest we came to nuclear war was the Cuban Missile Crisis, which was about 51 years ago - that really scared national political leaders.

I happen to think we have a pretty good international regime for preventing proliferation. When a country like Iran tries to get a nuclear weapon, there are all sorts of challenges. There are a lot of important things we should do to minimize the risk of nuclear conflict, but none of those things are moving away from nuclear energy. If the alternative is 100 dams on the Amazon or 100 nuclear power stations, most conservationists should prefer the nuclear power stations. If it is another coal plant a week in China, or a lot more nuclear, I think we should favor nuclear - China also already has nuclear weapons.

You have also been critical of a sort of apocalyptic environmentalism, which speaks in terms of saving the earth from impending ecological doom. A lot of us in the environmental movement got started with this idea that we are going to "save the earth." What are the consequences of environmentalists speaking in such terms, and under what circumstances should we take these concerns seriously?

That is a great question: obviously that was our motivation as well. There is obviously something really lovely about it. It demonstrates a real idealism and a real commitment to making the world a better place. And I think that is positive and should be encouraged. But your question gets at a dark side to it as well, which is a kind of Messianism, and even a kind of narcissism, that suggests we are going to save the world from all the bad people and from all of the disruptive sides of human-kind. And I think it has really fed a lot of the political tribalism and polarization that has actually made it harder to get things done on the environment.

When you talk to Republicans and conservatives, it is a remarkable thing: they do not think of themselves as hating nature. In fact, they quite resent being told that they are nature-haters and that they do not care about the environment. And in fact, I think it is quite bad for the discourse, I think it is bad for all sides in these arguments, for one side to be accusing the other of not caring about the planet. We should start from a place that says we all care about the planet and should be flexible about our solutions for saving it. One of the great ironies is that the two technologies that have done the most to replace coal or prevent coal power plants from being built, are natural gas and nuclear, and the party that loves natural gas and nuclear the most is the Republican Party, while those who have most strongly opposed those technologies are the environmental community and progressives. That should give us pause about who is actually speaking for the environment.

This is really painful to hear.

As a liberal-Democrat, it is sort of shocking to think of it in those terms. But the opportunity in it is that we might actually have a lot more common ground with people who are routinely called "global warming deniers," and against nature, and against science, than people previously realized.

That's beautiful. On another note, you have written about how environmentalists in their effort to save the Amazon Rainforest have failed to listen to the needs and aspirations of real Brazilians, and in doing so have caused significant resentment, and this has thrown up roadblocks to cooperation. How can environmentalists save the rainforest and at the same time support Brazilian developmental aspirations. And what might this look like?

The interesting thing in the years since our book came out is just how much of the Amazon has been spared by new agricultural technology. In particular, the thing that was happening, and may still increase, was the expansion of soy plantations in the southern Amazon. These are really different kinds of plantations than the older feudal-landlord, low-technology, and low-capital investments of the earlier model. These are high-tech, intensive farms, in the southern Amazon region, kind of a scrubland region, which they were clearing for soy. There was a big reaction from environmental groups against it, understandably.

And then something amazing happened, which is that the deforestation massively slowed down. I do not know if it exactly came to a halt. But part of the reason for that was that they were able to produce so much more soy on so much less land than prior farms were able to do in Brazil. This really took the pressure off of other parts of the Amazon. Certainly it took pressure off of other wild areas in the world. So, we were surprised by this as much as anybody. In some ways it is reminiscent of the shale gas revolution, which sort of comes as a surprise, even though it had taken decades to make happen. In the case of Brazil, the government was supporting research and development to change the soils, because they ended up using, I believe, phosphate to treat the soils for the soy plantations, and they had the big breakthroughs sometime in the eighties or nineties. And they had huge soy-production and beef-production, driven heavily by demand from China and Asia, which has been a major economic benefit to Brazilians, who have seen big increases in their standard of living and their quality of life, with people living in the slums and the favelas getting electricity, the government providing far more generous welfare benefits for the poor than they had previously provided in prior decades, all of which you can trace back to these technological innovations.

That being said, Brazil is still a highly unequal society, arguably more unequal, as a consequence of wealth. In other words, even as the poor get wealthier in many countries, the rich often get more wealthy, especially as you go from an industrial to a post-industrial economy. In the case of Brazil, what is happening now is that there is a massive popular unrest that is directly tied to the rising expectations. Most people forget that you often get protest movements in a period of economic disappointment that follows a period of high economic growth and higher expectations. In some ways that is what was happening in the sixties, and I think that is what is happening to some extent in Brazil.

So, the lesson that comes out of Brazil is really a story of a developing country rising, just like China rose, by intensifying agricultural and energy production. And it is now seeking ways to be more environmentally benign and beneficial. And that is the big story we will be writing about in our book, the rise of Brazil and China and the rest, in contrast to the west, using technology and becoming more green and more ecological as a consequence.

The idea that agricultural intensification, yielding more crops per-acre, through what is really conventional agricultural development, could do so much to save the rainforest, really throws a monkey wrench into the whole environmentalist narrative.

It is a very subversive idea, but unfortunately, or fortunately, the facts bare it out. We are just examining that forest data right now, and part of it is the exact same dynamic as we had at the TVA, as we talked about earlier. When farmers have electricity to power irrigation systems and pumps and all the rest, and you have tractors and you have fertilizer, you can just grow a lot more food on the same amount of land.

Malawi, in Sub-Saharan Africa, is an interesting case. The government had been told not to subsidize fertilizer for its poorer peasant-farmers, whose soils had been depleted of nitrogen and needed fertilizer. Government then went against the wishes of the IMF and the World Bank and all of the western economists. They subsidized the fertilizer, and the people loved it, because they had a lot of crops. We in the west kind of go, "Oh, it is conventional agriculture and the people of Malawi should have organics." It is like what you are talking about exactly, because they have organics and they are some of the poorest people in the world. And so the idea that they should not have fertilizer, or they should not have electricity, and centralized power generation, from either a hydro-plant or a coal-plant, is obviously not only hypocritical for us to be saying, it is actually a way of trying to deny development.

And I think that is a really important thing to come to, because if that is the case, and I think it is, then what is the environmental project about? It raises an existential question: "Is the environmental project going to be about actual development for the poorest people in the world?" And if it is, then it is just a farce to say it is all going to be solar panels and microchips and organic farms. That means you have a whole set of other questions that come up, like all the hardship and negative consequences of societies going from peasant-agriculture to large-scale farming and big cities and centralized grid-electricity. You have a whole bunch of consequences, many of which we are familiar with and went through ourselves. The transition from the countryside to the cities is quite wrenching. The Chinese are in some ways doing it in a more humane way than anybody else. In Brazil or certainly in the United States during the dust-bowl, it was a lot of poor people being hungry and displaced in slums. In China they are actually trying to move peasants into cities, but that is a direct consequence of increasing agricultural productivity and reducing the need for labor, for replacing labor with mechanical and chemical energy.

So, you get a whole set of other consequences that the environmentalist movement has not really had a chance to think about, because it has been so focused on saying it is going to be a different kind of development for all those poor people. They are not going to make our mistakes. That has been the line from the west, and that is just all gone now. And it is gone now, in particular, because it is 20 years after the Earth Summit in Rio. And China and India and Brazil and South Africa have all been clear about what choices they are going to make. Now the Congo is saying it is going to build, with Chinese money, one of the largest dams in the world on the Congo River, which is the second largest river in the world after the Amazon.

So, the world has chosen a particular development path. And the environmental movement that starts in the sixties and is maybe at its high in 1992 in Rio is just gone forever. Now the questions around climate, around agriculture, around energy, are significantly changed.

It seems many environmentalists have misunderstand how population growth occurs, and it also seems like a lot of them deliberately misunderstand how population growth occurs. I have often found myself telling highly-educated environmentalists, who are very sensitive and intelligent people, that the poorest and most insecure people populate the fastest, presenting abundant data and reasons why this is so, and reminding them that development almost always tends to bring about declines in population growth - although the best way to do so is through the education of girls. While there are many real reasons for environmentalists to be concerned about development, these correlations concerning population decline are empirically difficult to dispute. But it seems like the idea that we could support the developmental aspirations of the poorest and hungriest billion people, and to even aid them in their development, in the service of both humanitarian and environmental goals, just grates against everything many environmentalists have developed in their thinking about reducing resource-use. How do you think these beliefs and attitudes impact the global environmental movement?

They are terrible for it. When we point out that environmentalists are basically opposed to modern development for much of the poor-world, people say, "That is a straw man, what are you talking about?" And we say, "Then they should have grid electricity, even if it comes from a coal-plant or a hydro-electric plant. They should have capital-intensive, technology-intensive, farming." Then they say, "Oh God, that is terrible industrial-agriculture; that is going to be terrible for people." If that is your response, then don't go saying that it is a straw-man, pointing out that you are against development, because that is development.

You can be more-or-less green and more-or-less humane. But I do not think there is anything environmental about trying to keep people in organic-agriculture, relying on strictly renewable resources, which is mostly wood and dung. And I do not think there is anything liberal or humane about that. When American environmentalists, who are really liberals, understand that these are actually the choices that people have, most of them do side with the need for the poor to develop. But I think the apocalyptic pronunciations that the world will come to an end if the world's poorest people consume energy and food like we do, and produce it like we do, is reactionary Malthusianism, and there is nothing liberal or humane about that.

I want to highlight some of the most striking things I have heard in this interview and experienced through your writings. One is this in-

credibly strong emphasis on economic development, technological development, supporting human development, and how unusual this is to hear from an environmentalist, but also the way this economic and technological development can support a more environmentally sensitive and sustainable life for people. What is striking in all of this is how much these shifts would allow the environmental movement to feed off of the human aspirations that are already there, to unleash something, some creative potential, and how much easier it would be to build a movement around another movement that is already happening, a movement we do not seem to be able to hold back, that is in many ways the whole thrust of human world historical development; and that environmentalists have been viewing themselves with a finger in the dam, and that actually unleashing that breakthrough energy of human developmental aspirations could allow us to very easily unite with forces we had previously been opposed to and bring about the kind of world we want to see; and that world might actually look a little bit different to us if we were to let go of some of our old paradigm of environmentalism. I am wondering if these things capture some of the essence of what you are getting at.

We should just focus on the right things that matter for the environment. So, in the United States, right now, we have a lot of gas. We have enough gas that we could basically replace all of our coal consumption. And Obama, to his credit, is pursuing that strategy. But environmentalists should be more singularly clear and focused on the idea that what you really want is gas that is produced increasingly better, with continuous improvement of gas operations at every level - methane leakage, trucks, everything. Basically you want it environmentally clean, and you want it cheap, so it can replace the coal. And then you want coal to be a bit more expensive too. You want to make coal more expensive, which is what the present air-regulations may do. And that says very clearly what our priorities ought to be environmentally. It says that if you have got fracking operations in your county that are really negative, where gas-guys are just dumping the waste right there on the land, and not properly disposing of it, or their cement jobs are bad on their wells, then they should be dinged. And if the industry is not going to stop that, then the government should. And we should understand that gas is better than coal, at every level.

And then you kind of go, "Alright, so then what is better than gas?" It could be nuclear, could be solar, could be wind, though it sure takes up a lot of space. There is a big challenge with wind and solar, but nuclear obviously has its challenges too. Then we should be pushing the alternatives as hard as we can. Our focus ought to be making those technologies much better and finding ways to do that and being agnostic about what it is. That is a really different looking environmental movement than the one we have. Globally it says, "How are we going to work with China to make the next generation of cheaper, cleaner, better energy and healthier, cheaper, and easier to produce food-crops?"

You are guiding your technological development, your technological innovation with values that are really widely held by basically everybody - Republicans, Democrats. They want their energy sources to be cheap, they want them to be clean, they want them to be reliable. Those are the goals. Those are the values that should guide technological progress, and we should stop being so dogmatic about it. And we should understand better that innovation comes from collaboration, from really deep relationships of trust between people that work in government, people in the private-sector. And you actually do want a revolving door in the case of innovation, even if you do not want a revolving door in the case of regulation. There is a whole different environmental movement that gets constructed around this that has the potential to be widely popular, bi-partisan, and just much more life affirming than so much apocalyptic and Malthusian environmentalism.

FRANCES MOORE LAPPÉ

Frances Moore Lappe is the author and co-author of 18 books, including *Diet For a Small Planet*, which has sold over 3 million copies worldwide. She is the co-founder of Food First and the Small Planet Institute, a collaborative network for research and popular education that seeks to bring democracy to life. She has been a visiting scholar at the University of California, Berkeley and the Massachusetts Institute of Technology. And she is the winner of the Right Livelihood Award, also known as the Alternative Nobel Prize, "for revealing the political and economic causes of world hunger and how citizens can help to remedy them."

Her most recent book, *EcoMind: Changing the Way We Think to Create the World We Want,* won a silver medal in the Independent Publisher Book Awards in the Environment/Ecology/Nature category. In it she argues that our capacity for getting things done is impeded by several "thought traps." These thought traps frame our understanding of environmental challenges so as to produce fear and hopelessness. But we can break free of this inner climate of despair by seeing opportunities and learning to think like an ecosystem. Thus, we can bridge the gap between "the world we long for and the world we thought we were stuck with" through a series of "thought leaps" that might transform our lives and world.

THEO HORESH: *You have written extensively about some of the ways thinking on environmental challenges become what you call "thought-traps." What is a thought trap, and how do thought-traps hold us back from thinking effectively about environmental challenges?*

FRANCES MOORE LAPPÉ: A thought-trap is my term for a negative piece of our mental map. Literally, for a lot of us, we cannot see what does not fit inside our map. We see what we expect to see. And that is okay, our mental map is life serving. But my thesis is that the dominant mental map today, determining what we can see and what we cannot see, is fundamentally life destroying. And at its core is the premise of scarcity. The thesis of *EcoMind* is that we now are living in cultures trapped in a mechanical worldview in which reality is characterized by separateness, stasis, and scarcity. We are separate from one another in a ceaseless battle over scarcity. The first and overarching thought-trap is the idea that there are not enough goods, meaning everything from energy to food, nor enough goodness in human beings. There is more than one version of the scarcity scare, a Rightist and a Leftist version. But they come together on the premise of scarcity. Often the Right is associated with this idea that "Things are scarce, so we just have to produce, produce, produce." And the Left says, "Things are scarce, so we have to cut back, cut back." I believe both can be misleading.

We have hit the limits of the destruction and waste humanity can inflict without horrific loss and suffering. But that is very different from what is often heard: that we have hit the limits of *nature's capacity to meet human needs*. In no way is the latter true: think of the extreme inefficiency built into our food system: Three-quarters of all agricultural land is funneled into producing animal foods but they supply only 16 percent of our calories. Identifying nature's limits as the constraint diverts our eyes from nature's capacities. Plus, one in every four food calories that is produced is literally wasted. And that does not include the calories "wasted" when people eat more than their bodies can healthfully use.

It seems like a lot of what you are doing in Eco-Mind is bringing people's thinking back to balance. You are capturing the opportunities and the challenges within the framework you have set up.

And the key is a very simple shift from a framing, "We have hit the limits," to a framing of alignment: aligning with nature and with human nature so our needs are met as natural systems thrive. Currently we are creating societies perversely aligned with both. As we align our economies with nature's regenerative laws, and acknowledge both the negative and the positive in our own nature, we can create societies to bring forth the best in us. We can see a future we want. We can take a deep breath and say, "Okay we can do this."

And that is the challenge. When people hear that "We have hit the limits of nature," it is likely they think, "Oh, we have gone as far as nature can take us, so we have to go for genetically engineered seeds and geo-engineering." "We have hit the limits of nature" ends up backfiring. If people believe there is not enough, they become more possessive. I feel strongly that we need metaphors about aligning with nature. We can explain and show that our economic rules and institutions are now violating nature, destroying its generative power, and we can progressively align with nature to meet human needs.

We have interviewed George Lakoff, and he tends to lay out carefully constructed linguistic frames on single issues, which he builds up into interlocking systems of frames. But your project looks more like a massive reframing within ourselves, so that we come back into alignment.

The task is very much an internal reframing, as well as a reframing of the issues - absolutely. One of my favorite chapters in the book is the thought-trap about human nature, for example. The chapter argues that how we view our own nature is foundational. I tell about an experiment, so simple but so telling. Psychologists gave a group of women expensive sunglasses and then told half the women that they are not really expensive sunglasses; they are really knockoffs, really fakes. The two groups were then given a math test and asked to score themselves. And guess what? Those women told that they were wearing fake sunglasses cheated a lot more in scoring the math test. And my theory of what was going on here? Those who thought they were wearing fakes, thought, "I am already a fake, I might as well cheat."

What we believe about ourselves matters enormously. That is why I spend a lot of time in the book exploring evidence for what brings out the best in us: what brings forth our pro-social qualities, now identified through new anthropology and neuroscience. We are learning that humans have deeply engrained capacities for cooperation, for empathy, for shared intentionality. And yet we also know, like any organism in an ecosystem, that we respond to context. And so whether those qualities are expressed, or our capacity for brutality comes out, depends on social context. From there, and this is another sweeping reframe in the book, we can let go of finger pointing, because it has to be circular, coming back to ourselves.

One of my favorite lines in the book is, "If we are all connected, we are all implicated." And we then say, "If Wall Street did what they did, and brought down the whole world economy, where was I, why was I not paying attention when the rules were removed that allowed the crisis to happen?" It is a much more hopeful way of seeing, because, as I argue in another book, *Getting a Grip*, if we really think that it is the "bad guys" who are the problem, and we have to get rid of all the bad guys, that is a pretty endless process: new ones are being born all the time.

But if you take the eco-mind idea, that we reframe our own nature, then we can align our own nature with the conditions that bring out the best in us. These include the continuous dispersion of power, transparency in human relationships, letting go of the blame-game, and accepting common, mutual accountability. We know that under those conditions the best in us is likely to show up a whole lot more often. And what a relief that this approach does not require changing human nature.

So, how can human nature be used to bring us together to meet the climate challenge?

We drop the scare tactics, we drop the guilt tactics. They may have short term payoff, but they do not create a sustaining way of being in the world. What is missing now is a really passionate, positive, cross-issue movement on these themes of alignment. So, many of the messages of limits and the finger pointing keep us from uniting in a common mission. Both conservatives and progressives get it that democracy has been hijacked by private interests. Yet there is so little articulation of that, so little attempt to bridge the differences and come together. And we can come together, because 90 percent of Americans agree that corporations have too much power.

You talk about transforming, "something so frightening as to make us go numb into a challenge so compelling that billions of us will eagerly embrace it."

We have an incredible opportunity to get it right, and getting it right is what I call "living democracy." It is not just about getting rid of the next bad leader, but it is a different way of being and understanding democracy, as a culture, as a way of life. Living democracy creates opportunities for regular people to come together and reason together, and there is much evidence it works, but this process is still invisible in the popular culture. The greatest under-appreciated human need is the need of voice, of power, of dignity itself. And recognizing that need for dignity, meaning the sense of voice and agency, needs to be fundamental to all of our environmental messages. It is so much more powerful and effective than just saying, "You greedy person, you are to blame." Of course you want your children to thrive, of course you love nature and you appreciate other species. We need to be asking how we can all have a voice to enable the future.

Meeting the climate challenge may involve changing a lot of rules, and making some new ones in business, in government, and in our personal lives. How might we make the most of this process of rewriting the rules of social existence?

First, we have to believe that it is possible. Too many people seem to have given up on democracy. Compared with other democracies, the U.S. is an outlier in terms of the extreme way we have allowed private wealth to overwhelm public decision-making. One ranking that looks at how much private interests control political decision making in democracies measures their freedom from corruption on a scale of 1 to 100. And while it places Germany as the best at 83, the U.S. scored 29 with Tajikistan.

As democratic participation has eroded in the United States, it is still on the rise throughout the world, and supporting this also seems to be a part of your project.

Yes, definitely. What is really working to end hunger is the regular people who are creating social movements of non-farmers and movements of farmers that are transforming agriculture into an ecologically sound practice. This is working to change relationships, not just to the soil, but to each other, so it is happening, but it is not as visible as it could be, and that is why I feel so strongly about my mission as a writer and a public speaker.

You have written that democratization within families, villages, and governments can, amongst other things, help alleviate world hunger through bringing about a redistribution of resources. Can you talk a bit about how some of your solutions to world hunger might also decrease greenhouse gas emissions?

We need to recognize the degree to which the food system is contributing to greenhouse gas emissions. There is a very credible study that says that the food system accounts for as much as 57 percent of greenhouse gases if you include deforestation, as you should, and concentrated livestock, and all of the transportation involved in growing food and bringing it to market. They are all related to our food system. To take just one example from the Rainforest Action Network, Indonesia is now the third-largest greenhouse-gas emitter because of the palm-oil operations, which are causing deforestation. We can also look at this through the lens of food waste: since about a third of all food is wasted, if food waste were a country, it would be tied with Indonesia as the third-largest greenhouse gas emitter.

Agro-ecology is better for the health of farmers, because they are not exposed to pesticides. It is better for consumers, because we are not eating harmful chemicals. It is better for the water, because it does not leave the water with nitrates and other harmful chemicals. But we are also coming to appreciate that through agro-ecology, through enhancing the health of the soil, and through creating more life in the soil, it holds more carbon. And so, agro-ecology is a very direct contributor to balancing the greenhouse gas imbalance.

Because of the release of carbon and the tilling of soil and the erosion of soil...

Agro-ecology reduces the release of carbon, as well as nitrous oxide and methane, very potent greenhouse gases, and it holds more carbon in the soil. Interspersing trees with crops can be especially helpful. In Niger, farmers have re-greened over 12 million acres of land through nurturing and growing 200 million trees. The trees supply fodder for livestock, provide fruit, and protect the soil from being eroded by wind and rain. So, agro-ecology and agro-forestry can make a very big contribution to righting the carbon balance, if spread to all of the areas in the world where soil is now degraded.

This was made real to me when I was in India 12 months ago and found myself in the midst of this incredible story of people who 20 years ago had been living every day at the absolute edge of starvation. They described it as a dark, dark time, in which they were bullied and beaten by their husbands and landlords. They described how they only had one Sari to wear, so they could not even bathe properly. They had to wash part of their Sari, and then let that dry, and then wash another part.

Then they shifted to agro-ecology. And I walked through this field of about an acre that had 20 crops in it, all growing nutritious millets, which typically have more than three or four times the iron than rice. They were growing oil seeds, they were growing protein in the form of lentils, the dal of India. They were growing greens and medicinal plants. They had everything they needed. They had completely created food security, village by village, in about 75 villages, in a very dry area. If somebody had shown me this land, it would have seemed so infertile, because it did not look dark and deep like ours but kind of red and dry. And I said, "Are you worried about climate change?" They said, "No, we know our seeds." They save seeds and share seeds and know which ones to plant, even if the rains do not come. So, I just saw a wholly different way of thinking about productivity, because there was virtually no waste. And they came together and created their own market, so that they could sell what they did not need; and together bought their own simple processing machine, so they did not have to process everything by hand. They had their own radio station, so they could tell people about it - it was all working.

When people say, "We have enough food to feed the world easily, it is just a question of distribution," this can be a sort of conversation stopper. And if you press into this answer, they seldom have a plan for distribution. It is interesting to me that you do not see redistribution as some great aid program for the world. It is not even so much a question of what we are doing with rich-world agricultural subsidies. The redistribution occurs at the level of the poor peasant family, at the level of the village, at the level of the state – say, the state of India or the state of Nigeria.

Exactly. People often say to me, "What you are really saying is it is a distribution problem." They seem to think I am saying we need a better global catering service. No, you just nailed it, what you said is absolutely right. Virtually every part of the world has the capacity to grow the food they need. But, don't misunderstand: I am certainly not against all trade, per se.

Think of India, where they have the most hunger in the world. Almost half of India's children are physically stunted, which impairs people for life. Yet, in the summer of 2012 India had a 71 million-ton stockpile of grain. And we calculated that the stockpile could provide one-and-a-half cups of grain for every person in India for the entire year. That was just the stockpile. So, clearly the challenge is not lack but enabling all of us to power and dignity. Dignity captures the idea of voice, of power, and of participation, whether at the level of the village or right here where I am today in the Boston area.

If one has power, one can eat, and if one does not, then one cannot eat. So, that is the fundamental question. And then there is the question of how we come together to create sustainable production in every sense of the word, because yes, there is enough food in the world, but a great deal of it is grown in a way that is not sustainable. The main idea is to shift away from the frame of "them apart from us," that they do not have, but we have, and we need to ship it to them. Once we see our real needs are aligned, we see can ourselves as part of a global movement, standing alongside one another shoulder-to-shoulder.

And this fits very easily within your general framework of Eco-Mind, insofar as this is an empowering message to these poor villagers and families in the developing world. And it can be an inspiring message. It also follows trends in democratization that are happening organically in the world right now, where the number of democracies is still rising in a stair-step fashion, and where we have seen a virtual explosion of pro-democracy movements arise in recent years. But it also seems difficult to effect from the rich world. How can we have an impact on this deeper democratization of families and villages and states to meet basic food needs?

If *"we"* is understood as ordinary citizens, who are changing our consciousness in order to connect directly with people in the global South, that is already happening on a major scale; whether it is my trip to Andhra Pradesh, India last year, where I got to visit with the Deccan Development Society and then got to tell their story, or whether it is movements like La Via Campesina, a global small farmers movement, creating a collective voice to change trade rules that make it harder for small farmers to access markets for their products. There is more and more cross-cultural, cross-national solidarity.

With your 1971 bestseller, "Diet for a Small Planet," you became one of the earliest environmentalists to promote eating less meat. This is a very intimate issue for many people, who struggle with the ethics of whether or not to eat meat. How has your thinking on this issue evolved?

It began with this feeling of, "Wait, we don't really need meat?" Because I grew up in cow-town USA, in Fort Worth, Texas. And it was just assumed that you could not live without meat. So, in the 1960s, when I first started looking at this, it was just such a shock to realize that plant foods would do us just fine, thank you very much. And now there is so much more knowledge about the health benefits of plant-centered diets. So, at the beginning it was just like, "You mean the experts are telling us we are running out of food, but actually we are feeding so much of it to animals that we shrink it to a small percentage of what can nourish us? That does not make any sense."

The initial impetus was just to share with people the idea that we are creating scarcity out of plenty and that this is not inevitable; there is no scarcity. It was just a very raw sort of wake-up. And then I rewrote *Diet for a Small Planet* in 1975 and tried to make even clearer the idea that a grain-fed, meat-centered diet is simply an outcome of a social-political-economic ideology that concentrates power, so much so that billions of people do not have the economic power to make *market demand* grain-to-consumer directly. Rather, the grain is made so cheap that it makes perfect economic sense to feed it to livestock and create what is a luxury product for many people instead. So, I tried to explain more clearly that the problem is not meat per se, it is what we can learn by studying the grain-fed, meat-centered diet. I wanted people to register that there is nothing inevitable about it, nothing good for our health about it, nothing good for the environment about it, nothing good for poor people about it. It was just something that was created by an economic system that returns wealth-to-wealth-to-wealth, until we have such inequality that today only 43 percent of the grain produced in the world goes directly to humans. And now we know the livestock industry is a huge contributor to climate change as well.

The change started with just with a raw wait-a-minute moment: we are creating scarcity from plenty. Eating a plant-centered diet became for me like putting a string around the finger, reminding you every time you eat that by choosing a plant-centered diet you are actually aligning with what is best for nature, what is best for your body, what is best for others. That is just such a positive motivation, and it keeps you healthy. Then in recent decades my focus shifted more and more to the climate impact of our diet choices, with my wonderful daughter Anna writing, "*Diet for a Hot Planet*." And then even more recently, the realization that our diets are increasingly providing calories without nutrition has added another layer of urgency about the healthy, diverse, plant-centered diet for our health.

It seems to be another one of those areas where so many things can come into alignment at once. We come into greater alignment with ourselves, we come into greater alignment with animals, we come into greater alignment with the environment, with hungry people. There are just so many issues that can be impacted through this one choice, with each supporting the others in a sort of virtuous cycle.

Not to mention the fun of creativity in the plant world. That is where the variety is, hundreds and hundreds and hundreds of choices in terms of fruits, vegetables, root-crops, nuts, seeds, all different kinds of grains. It is just endless varieties of things that you can combine to create wonderful meals, as opposed to sticking with meat in the center of the plate and plopping down the vegetables. I just find it so endlessly exciting to be a cook in the world of plant foods.

There is something about restrictions that seems, if approached in the right way, to bring about greater opportunities for creativity. I hear you saying this regarding diet, but I also see it in the explosion of clean-energy technologies.

I think so, though in principle I hate the word "restriction." But one can be locked into a frame, right? You grow up with meatloaf and potatoes and vegetables. That is supper. Then, if you just throw that frame out, what a "meal" is suddenly becomes open-ended. And so it does really spur creativity.

You are quite critical of quantitative thinking, and yet this seems to be the only way for both scientists and economists writing about climate change to think clearly about how minute changes today might add up over say, the next 100 years or so. Thinking quantitatively is also perhaps the best way to discover what will be most effective in reducing greenhouse gas emissions. By paying close attention to the greenhouse-gases emitted for heating buildings, driving, or bringing food to market, for instance, we can get a good sense of what to prioritize. How might we make the most of these quantitative approaches to thinking about the climate challenge while bringing to our thinking the kind of qualitative approach that allows us to also see the world through the lens of an eco-mind?

That is a great question. First, it is not either/or, because clearly I deal with the quantitative all the time. But what is the qualitative implication of the quantitative - that is the question? I just gave a speech at the World Food Prize event, where Monsanto and Syngenta were honored. My talk was a very direct challenge to *productivism.* I am in no way saying production does not matter; no, of course, we have to produce enough food. But how that quantity is produced and who has power over what is produced are relational questions. All of that involves seeing the relational implications of different types of seeds and actual farming practices. If that seed is patented, and it is controlled by an oligopoly, that in and of itself determines a lot of the relational consequences: dependency versus symmetry in human relationships. So, I am not in any way denigrating the need for quantitative analysis and for quantities of food. I am asking us always to put the quantities in a relational context, asking what are the relational implications.

How would you respond to a critic of a work such as this who says, "It is all well and good to change the way we think about climate change, but the real problem lies in changing the institutional policies that drive climate change. It does not really matter how we think about these problems, so long as we act."

I agree with that, of course, but fundamentally, our thoughts drive our actions. Depending on what we think, our actions will be very different - or we will not act. I could not agree more that ultimately it is what we do that counts. I also believe humans, to be happy, need to feel they can make a difference in the world. So, all of the re-framing I present in *Eco-Mind* is about freeing us to engage in useful actions and to put forth messages that will motivate others. Like I was saying, the message that we have hit the limits could cause people to think, "clearly nature cannot meet our needs anymore, so genetic engineering is the answer." The key for me is developing messages that encourage people to see the problem not as nature's limits but our violation of nature. We align with nature and there is enough for all.

And how would you respond to those desperate environmentalists, who think that it is already too late to save the planet from destruction?

It is not possible to know what is possible, so I do not describe myself as an optimist or a pessimistic, but as a possiblist. And I think if we are going to take up room on the planet, then we might as well recognize that the nature of life is continuous change. It can be glacial in its pace or surprisingly speedy, so staying in this place of possibility is my everyday challenge. And it seems to me most appropriate, given what we are coming to understand about the nature of life - that we really do not know. And all we can know is that if we do nothing, if we just continue on this track that is so deadly for us and other species, it is not going to turn out well. But our species has never been here before, conscious of a threat to its very existence.

I just turned 70, so I have now lived through an era in which there was no consciousness that we were creating our own demise. Now, for growing numbers, it is an everyday understanding that the path is undoing the very foundation of life. One of the things I do in *Eco-Mind*, is to ask readers to think about Germany, where I am going in two weeks. Hitler was alive when I was born, so for me it does not seem like that long ago. In my lifetime, Germany was *the* pariah in the world. It was *the* fascist horror. And yet today Germany is a powerful example of leadership at so many levels, certainly in terms of the environment. How did that come to be? Nobody would have predicted it. I would not have predicted it. That is humbling, and it is a thought that keeps alive my mantra that "it is not possible to know what is possible."

It is similar with the women I just met in India and the stories they told me about how they lived just 20 years ago. The transformation has completely changed their gender relationships. They say now that if a man beats a woman, the whole group of women go to him and confront him. While we were sitting there having our conversation, a young man brought us tea. Who would have predicted that? These are the lowest-caste women. Clearly, a sense of self can change radically, the sense of what is ones own capacity.

We have never been here before, ever. And we do not know what the consequences could be, how quickly we could break out of old patterns. In other times in life, we have seen how an enemy, in the negative sense of an enemy in war, can bring a people together. Natural disasters also bring people together. If we can transform the global poverty and climate-change challenge into an emergency breaking us out of our old differences and preoccupations and pettiness - as can happen in emergencies - then we do not know how quickly things could evolve.

PETER SENGE

Peter Senge is a systems theorist and the bestselling author of six books, who consistently appears on the lists of top management thinkers. In 1997, Harvard Business Review named his book, *The Fifth Discipline*, one of the seminal management books of the last 75 years. In 1999, Journal of Business Strategy named him a "Strategist of the Century," one of 24 people who have "had the greatest impact on the way we conduct business today." And in 2001, Business Week rated him one of the top 10 management gurus.

But Senge is at heart a systems theorist, and it is out of his work in systems theory that his management writing arose. Systems theory, as it is applies to organizations, looks at the relationships between individuals within a firm, between individuals and the firm itself, and between the firm and its physical and market environment. The firm is always a system of systems, enmeshed in wider systems. In taking this larger view, systems thinking highlights the constraints on organizational change and the leverage points through which deeper and more lasting changes might be made. This work is vital to climate change, because reducing greenhouse gas emissions will require extensive changes to business cultures. And it is notoriously difficult to make such changes stick.

Senge is a Senior Lecturer in Leadership and Sustainability at the Massachusetts Institute of Technology, Sloan School of Management. Senge is best known for the idea of the *learning organization.* According to Senge, learning organizations allow people to "continually expand their capacity to create the results they truly desire." His work suggests a path through which personal and institutional changes can be mutually supportive and through which climate change can become a catalyst for continuous learning.

THEO HORESH: *You have written a lot about systems theory, as have many other environmentalists. Can you say a bit about the significance of systems theory to your work and to sustainability work in general?*

PETER SENGE: Systems theory is a term that can sound pretty abstract and academic. But really it is all about understanding interdependence, and interdependence is life. We grow up in webs of interdependence, we just have little or no training in how to understand those webs. Family is a system, a network of friends is a system, a playground is a system. So, as kids growing up we are immersed in interdependence, social interdependence, but also biological and ecological. You could say all the challenges that they refer to as sustainability challenges arise from our not really understanding the interdependence or the webs of interconnectedness within which we do everything we do. Nobody is trying to create climate change, it is nobody's explicit goal. Yet we do that as a byproduct of taking actions like depending on fossil fuels, where we do not really understand the side effects or the unintended consequences of those actions.

How does systems theory change the way we might look at climate change?

There are basics that everybody needs to understand, basics in say, the system of how greenhouse gases like carbon dioxide accumulate in the atmosphere. Many years ago, there was a study done by one of my colleagues at MIT. This was done with Masters students at MIT and the University of Chicago, bright, young people. And two-thirds of them thought if you reduce the emissions of greenhouse gases, if you reduce CO_2 emissions, you would start to reduce carbon in the atmosphere. They had no concept of a bathtub. As you know, CO_2 accumulates in the atmosphere just like water flowing into a bathtub. They just had no concept of the very, very basics, the difference between the inflow and the stock. The bathtub level rises when the inflow is less than the outflow. Every kid knows this, but because we are not trained in understanding systems, we do not even transfer the most rudimentary understanding of the delays in this system. Not understanding these delays is a crucial reason why people take the otherwise very rational attitude of let's just wait a little while. Let's see how bad it is, and then when it gets really bad we can do something. But by the time it gets really bad, it may be very well too late to do much of anything.

Prior to the Necessary Revolution in 2010, I had read almost every book you had written, but I had known you as a writer on management and organizational development. What made you decide to write a book on sustainability?

Actually, it is the thread that has always run through everything. Sustainability is a contemporary term, but in my field, it is understanding complexity and systems. I first came to MIT as a graduate student and landed in the middle of a group doing their study on the *Limits to Growth.* All the things we now call "sustainability issues" were pretty much identified in that book. So, it is something I have grown up with.

And you have applied this thinking mainly to organizations.

Yes and no - I am part of a much bigger field. The man who did that study I was just referring to, of people's mental models about slowing climate change, built the first energy economy system model in 1975. So, there is a long tradition of applying these tools to understand larger systems, and that is, again, the tradition I grew up in. And even when you look at in a business, it is still the same basic challenge. How do people understand interdependence? How do they see multiple feedback loops to their shaping? How do they behave business-wise?

The best products do not make the best company. You have got to put all the pieces together - that is a system. And then how do we understand the implications of our current actions? And then what do we need to do to bring about change, particularly when we are dealing with these very complex webs with lots of delays and lots of unintended side effects? So, it is all the same stuff. The fundamental idea of the whole systems perspective is that we live in webs of human interdependence. The problem is a little bit like if no kid were ever given a musical instrument: then there would not be many musicians in the world. You may have latent ability in music, but it has to be developed. We have enormous latent abilities to understand interdependence, but for most of us it is completely undeveloped.

You have written, "All organizations, especially those that live for a decade or longer, have within them the capacity to create and innovate. The problem is that often this capacity atrophies. Pursuing the profound innovations needed for life beyond the industrial-age bubble offers an extraordinary opportunity to tap and even grow this regenerative capacity." What might this regeneration look like?

Everything we can imagine has to change. We have got the wrong products. We drive the wrong automobiles. We use the wrong devices, powered by the wrong energy, often created through the wrong processes. So, if you look at it from an economics perspective, the processes, even the business models themselves, are all not what you would look for if you really wanted to create a society that could live in harmony with a larger natural world. We throw away an incredible amount of junk. The typical estimate is that, as average Americans, we use about a ton of raw materials per day, almost all of which is waste in processes of extraction and production. Now again, nobody sees it. That is the web of interdependency we just do not see. No one in their right mind would think that is an effective way to run an economy. Most of us grew up with things our parents would tell us about waste, like "waste not want not." My mother used to always say that. It is a very old idea that you can find in most cultures, and yet the amount that we waste to support our everyday ways of living is enormous, so zero-waste would be one starting point.

A second starting point would be energy. Why do we think we can get our energy from a different source than every other species on the planet? Every species on the planet, almost every one, lives on one source of energy: it is the sun. Why do we think we are any different? We dig up that stuff, stuff that died millions of years ago, and burn it. And then we say we are really sophisticated people, but that is not very sophisticated.

So, you start with products, and say the products need to be all part of a larger material system, consumer waste. Then you go to energy, and say the energy has to come from the same places as the energy for every other species on the planet. But then when you really think about these things more deeply, you realize that all of that is being driven by a culture. There is no way that any of these things will be accomplished if you have a culture where people think that well-being is associated with rising incomes and rising material acquisitions. So, ultimately this is all about culture.

You have written about the efforts of several major corporations to become better stewards of the environment. What sorts of forces in these organizations tend to drive their sustainability initiatives?

There are definitely forces, but there are also people. The most compelling part of all the stories that we have been part of is that they are the beginning of the beginning. That is very important to recognize. The people who are really sober in the organizations who have made a lot of progress will say, "We have gone 1 percent of the way." We are shifting our huge ocean tanker here and it takes a long time.

That said, the forces arise from a little shift in the strategic context for that business. That shift is occurring for all businesses, but it is just not perceived by most for the same reasons mentioned before. We just do not perceive so much of the effects of our day-to-day actions.

Let me give you an example. One of the companies that has been a big part of all of the work we have done in global food systems is Unilever. Now, Unilever traditionally is not a very innovative company. If you go back 20 or 25 years, it was a company that really grew very, very slowly, like 1 percent a year. No one in their right mind would have called Unilever a big locus of innovation. It was not particularly attractive for young people to go to work in Unilever. In fact, most people had never heard of Unilever. You know their brands: you know Birds Eye, you know Lipton Tea, you know Ben & Jerry's. But until a few years ago, there was no brand identity for Unilever. Now all that has changed a lot, and the reason it has changed is because there was a huge wake-up that our global systems of food and water are clearly not sustainable.

We have lost half the top soil in the world in the industrial age, about 1 billion hectares in the last 50 years. Water is the most acute problem in the world. The World Health Organization estimates that the number of people without reliable access to clean drinking water could be about 2 billion by 2020. So, there were a few people at the senior levels of Unilever, who looked out and said, "We have two types of products: food products and health or beauty products. The global food system is a disaster, and two-thirds of the world's water goes to agriculture. A lot of the rest of the water is being used for washing things. We are right in the middle of this huge water crisis as well as the unsustainability of the global food system." So, they started to see the strategic context of their business was shifting. Now, this goes back about 17 or 18 years. I remember talking to the guy who at that time was the Chairman, and he basically said in very simple terms,

"If there are not fundamental shifts in the food and fishing industries, we just will not have a businesses worth being in within a couple of decades."

That is not philanthropy, that is not do-gooderism. That is just strategic thinking - looking ahead, and seeing if we continue the way we are going in the larger systems we are part of, the conditions for having a healthy business are deteriorating. The food industry has had a huge wake-up call today. Now we have a network of about 70 of the world's biggest food companies and NGOs all working together on this. It is an industry-wide awakening, but it all starts with seeing this shift in strategic perspective.

You have written a lot about snowball-effects and tipping points. Can you talk a bit about not just the snowball-effects of sustainability initiatives within a single business, but within a sector of the economy, and in the global economy as a whole, and how we might make use of the snowball-effects of other efforts that we are not involved in to spur on our own.

This "snowball effect" is another feature of systems. Everything in the living world that grows, and for that matter the growth of an enterprise, the growth of a business, arises because of self-reinforcing dynamics. Two cells, four cells, eight cells, sixteen cells, thirty two cells, this is the basis of the process underlying the growth of every single organism. And it is very similar to how a business grows. You have a few customers, you do something that really excites them and makes them want to do more, they let other customers know, and before you know it, you have got a snowball-effect going in the growth of customers and a business. Everything that grows in nature has this kind of process sitting behind it - it is nothing that we do not already understand.

Now it can be positive or negative - the threat of a rumor is a snowball-effect. But in most cases, most of us find this threat of rumors to be natural, and we really do not think it is too terrific, because it can reinforce very distorted views. So, that is the basis of a snowball effect, and anyone who is interested in any big change really needs to always ask themselves how it could become self-reinforcing, which is a big problem for a lot of *change agents*. It does not matter if they are working in an individual organization, or as you said, trying to create a real shift in an industry or sector of our economy. A lot of times people have a real definite idea of what they think needs to change, but they do not have a very definite idea of how it can become a process of self-reinforcing growth. If they do not have that, it is a big problem, because then they end up just having to keep driving the change.

The environmental movement has been replete with bad strategic thinking. We are going to tell everybody how bad something is, that we are destroying species or whatever. People are going to see this as really bad. They are going to get really afraid that this has got to change. And then, by golly, they are going to rise up and do something. But there is nothing that reinforces that for them. You get short term episodes of concern and engagement, and once people think things are a little better, the energy for the change goes away. We have made a lot of progress on a lot of local pollution problems in the United States. And that has always come, because of some sort of movement to alert people that we are destroying our river, or we are polluting our air or whatever. But you get an episode of engagement followed by things getting a little better, and then people go back to business-as-usual. There is nothing that makes it self-reinforcing. It is really the first principle of applying the systems perspective to change. It is not just about changing things - that is not enough.

It also seems like this cycle of alarmism, followed by strong action, is for many people also followed by burn-out.

Exactly, that is what I was alluding to when I said the change agents have to keep driving the change. That is a terminology which is really crazy. "We have got to drive change." You do not drive people, you drive an automobile. It is a very mechanical image, and the side-effect of it is the people doing the driving are continuing, and they develop a mindset about this that they are producing the change, rather than eliciting processes that will become self-reinforcing. If you think about growing any business, it is not about going out and getting 10 customers. That is great, you may need to do that initially. But then it is those customers getting other customers that really grows the business, and that becomes self-reinforcing.

You have also written about the need for teamwork and collaboration to drive change within organizations, in businesses, and in non-profits. Can you say something about the significance of teamwork and how it is best achieved?

This is another way to talk about the systems perspective, which is very helpful, because again, systems sounds very abstract to people. It is all about collaboration. Shifting a larger system requires a critical mass of the people in different places in that system to start to have a new vision, to start to have new ideas that orient their actions, to do things locally that accumulate into larger impacts.

It is always about collaboration, and there are always at least two levels of collaboration that are really important. There is a group of people, who are actually working pretty closely, who we often call a team. We use the term team to mean any group of people who have to work together closely to get something done. But then there are always larger networks too, people you do not know, people you do not have direct contact to.

Those two fundamental systems, you might say social systems, teamwork and network, are always what is behind any significant scale of change. That is not just a euphemism or a nice way of talking. It also implies certain skills. Again, a lot of change agents have great ideas of what needs to be changed, but they do not actually have a lot of skills they need to pull it off, and one particular area of efficiency is often collaboration. They are often not very good at listening. You do not build collaboration, either in a team or in a larger network by people who cannot listen to each other very well. They are so busy advocating their point of view. They are not very good at listening to others, and seeking together, coming up with a view together, coming up with a way to synthesize different people's points of view into a larger strategy. Those are really key skills, which are often just not present in a lot of change agents.

Say more about the personal qualities that are necessary not only for collaboration, but also for sustaining the change-work.

We always say there are three domains of core leadership capabilities. It is kind of a foundation of all of our work, and one of them is what I was just referring to, which is the capacity to collaborate across boundaries, particularly people who are not like you. It is great to put together a team of a bunch of people who all think the same. The problem is that will never represent the larger system, which you are trying to move. You are going to have to go collaborate among people, who are quite different if only thinking different agendas, different histories, different cultural perspectives, and different organizational perspectives. So, that is one domain, the capacity to build collaborations across all sorts of boundaries.

The second is what I have been referring to since the beginning of our chat here, which is that ability to see the larger system. If you cannot answer the question how this process will become self reinforcing, then you are not seeing the larger system. And as we were saying a few minutes ago, you can end up with a lot of episodic change efforts, not something that really grows to a scale that will really make a difference.

The third core capability is the ability to help you and others shift from just seeing problems to seeing real opportunities to create. The energy for any change process really shifts when people go from just reacting to problems to seeing a real opportunity to innovate. It is what we were referring to earlier. The energy of innovation is really all about creativity and creating something new that is really of some benefit to somebody. The energy of innovation is really different than reactive problem solving. Again, if you look at the environmental movement, or for that matter the social justice movement, you see this same change strategy again, again and again. "Hey guys, we have got this big problem, this is awful. We have got to do something about it, let's go fix it."

Typically, we are pointing fingers at some of the bad folks, who create the problem. It is full of reactivity and demonizing and it just does not produce sustainable change. We all know this personally. You know if you are just reacting to problems, because your life is just one problem crisis after another. As you were saying before, pretty soon, you tend to get pretty burnt-out. People who are creating something new, or who are entrepreneurs, or who are building an enterprise, they have got a lot of problems. Of course, life is always full of problems. It is not that problems go away, it is that what is predominant is not the problem but the excitement of pursuing an opportunity. That is the, you might say, psychic or psychological or emotional energy of innovation. So, shifting from reactive problem solving to innovation orientation is absolutely crucial to all leadership. Again, it is not an either/or, it is just what is predominant.

I am struck by the numerous levels of health you are addressing.

There is the health of the individual, the health of the organization, the health of the political and economic systems. It seems you are not just looking at these big systems: you seem to be very alert to the role of individuals within teams, the role of teams within organizations, the roles of organizations in the economy, and it just goes on and on, these nested hierarchies of relationships. Can you say a bit about the alignment of all of these healthy processes?

It is not that easy to even formulate the key questions here, but it is a little bit like what I was referring to awhile ago when I said this is really about our culture. It is about how we live on all these different levels, how we live as individuals, what orients us, what are the things that really give meaning to our life, what are the sources of our personal health and well being. It is the same in a family, the same in a team, the same in an organization, the same in industry, the same in society. It is the same basic question again and again. What are the sources of our well-being? What do we really want to conserve? What I mean by conserve here is, if I look at any of those levels of systems, there are things that we really like. It is not like our entire lives are a disaster. We have a lot of relationships that mean a great deal to us. We really want to conserve those.

From the standpoint of the strategy of leading change, there is a subtlety that people often miss, because they are often so focused on what needs to change. They simply forget the question, what do I not want to change? What is it that we are already doing that is in line with what we want to create? We are very healthy in a lot of ways, but we do not notice the things that make us healthy. We do know how to pause, we do know how to relax and how to attend to our relationships to some degree. And we always have a lot of areas where we can improve, but that, again, is what I was referring to as a shift to the creative orientation, because we are already creating a lot of what we would like to conserve. There is a lot of other stuff that we can let go of, and if you take that attitude, it really starts to become a powerful line of inquiry.

It is very hard to break a bad habit. I think I need to buy this thing, but maybe I do not need to buy it, maybe it is just crazy consumerism, and I have been programmed by the consumerism. That is what it is like to break a habit, it is very difficult. So, the way you break a bad habit is you start doing something new that becomes more habitual, and eventually more powerful, than the bad habit, right? Stopping smoking is really difficult, and people try to stop but keep taking it back up. But if you look at processes that work, it is because they get into walking or they become part of a group that supports each other, and the meaningfulness of the relationships in that group actually becomes an important part of their life. They are starting to grow something new, rather than just getting rid of something they do not like.

It is that whole orientation of what we want to conserve. What is it that we know is meaningful to us that we know how to do on a limited scale, but about which we are really confused? Another way to say all of this, when I talk about the cultural changes, is we are simply confused. We really have, in one sense, a good understanding of what matters, but if you actually look at how we live our lives in total, we are pretty confused. We do get manipulated by advertising, we do get manipulated by our fears, we do get manipulated by views of ourselves which are very self-limiting, as opposed to just noticing the things we actually do well already and want to conserve and grow. So, yes, there is an orientation here that implies all those different levels.

There is a common theme that has continued to arise throughout The Inner Climate. As Mike Hulme puts it, "We should be asking not what we can do for climate change, but what climate change can do for us." It is this theme that climate change is an opportunity to re-evaluate everything we are doing on the planet personally, organizationally, politically, economically and internationally. Can you talk about the potential for regeneration climate change presents?

Climate change is like a 40-year-old stroke or heart attack. People who have had that sort of event occur to them or people who have had cancers that they have encountered in the middle years of their life, will in retrospect talk about the gift of their stroke or the gift of their cancer. Because it caused them to wake up. It caused them to start thinking much more deeply about what the hell they really care about. Climate change to me is like that.

There are so many aspects of the unsustainability of the global industrial process. You can look at the destruction of species, you can look at the destruction of the oceans. Climate change is just one shift in a whole cluster, but it is one that gets harder and harder to ignore. Even people who are skeptical about the human sources, or the human influences, on climate destabilization, can see that the weather patterns are getting bizarre - the pervasiveness of extreme storms, typhoons, hurricanes, etc. If you pay any attention to the local critters in your neck of the woods, you notice the birds that are gone, the animals that are shifting. People can see all this, so it is getting to be pervasive destabilizing to our everyday normalcy.

We take a certain predictability in climate as a given. It does not matter where you live. If you live in Chicago and your weather has started to be like the Yucatan, you would feel very strange. But if you lived in the Yucatan and the weather started to be like Chicago, you would think it is very strange. This strangeness is starting to become something that people encounter again and again. Now, the downside is human beings can learn to renew themselves, or ignore all kinds of things, right? That is why a lot of people have a significant health event in their forties, but they do not change. They learn to adapt, "It is just one thing, just one event." Some go back to the way they were living, and others wake up. Climate change is potentially a wake-up if we use it that way, because it is really telling us that our natural environment is changing. We have all these signs that things need to change, but climate change is one that cuts across and influences all of them. It is kind of the pervasive heart attack event for human kind if we pay attention to it.

SECTION II

PAUL SLOVIC
ANDREW REVKIN
GEORGE LAKOFF
JULIET SCHOR
PAUL EHRLICH

PAUL SLOVIC

Paul Slovic has been producing groundbreaking research on how humans experience risk for half a century. He is the co-author or editor of 11 books, including *The Perception of Risk, The Feeling of Risk, and The Social Amplification of Risk*, all fields in which he is a pioneer. And he has published over 300 papers and chapters on decision making, risk perception, risk communication, and other related fields.

His work is regularly referenced by climate thinkers, and it lies beneath many of the quandaries presented in this book. The human experience of risk is less related to actual dangers than to our ability to imagine their consequences. This means that risks like global warming, which are difficult to visualize, tend to be experienced as less risky than say, terrorism or death by cancer. Slovic has also pioneered research into how psychic numbing occurs in mass tragedies, like genocides, demonstrating that the greater the number of people affected by some event, the less we will tend to feel for them. This is particularly relevant to climate change, where such massive numbers will be affected.

Slovic is a professor of psychology at the University of Oregon and the founder and President of Decision Research. He received his Ph.D. in psychology from the University of Michigan and he is a past President of the Society for Risk Analysis. In 1993 he received the Distinguished Scientific Contribution Award from the American Psychological Association. And he has received honorary doctorates from the Stockholm School of Economics and the University of East Anglia.

THEO HORESH: *If all goes as expected, global temperatures will probably rise by something like 3 to 7 degrees Fahrenheit over the course of the next century, but there is a very small chance that temperatures will rise 11 degrees or maybe even more. How do people tend to respond to these sorts of probabilistic scenarios?*

PAUL SLOVIC: The main response is to do nothing, to change very little in our behavior, not to treat it as real. The question is too remote to answer. I think the numbers, 3 degrees, 7 degrees, 11 degrees, are not meaningful to us. We know what a 3 degree increase in our current daily temperature means, and it is really nothing that we cannot easily adjust to - we hardly notice it. We just do not have an experiential basis for understanding what these numbers mean to the climate and the environment.

How do you think the world would respond if global warming pollution were killing a single beloved individual, say the Dalai Lama, and the whole world had to watch him dying day-after-day in his weakened and vulnerable condition - but we could save him if only we could pass a global climate deal?

You are right to think about how a singular, individual dramatic incident can do a lot to raise awareness, at least for a short period of time. We see that, for example, with regard to deaths from gun violence. It may not happen all the time, but when an event has special qualities, like the shooting in Connecticut the winter before last, where 20 young children and some adults were killed, then suddenly we sit up and take notice to a problem that has been around us for a long time.

And your research has also indicated that the smaller the number of people involved in these sorts of tragedies, the stronger our response might be per person.

In general, we can understand and relate to, and make an emotional connection to, harm to an individual, particularly if we know something about that person. When the harm comes to multiple people or masses of people, who are represented by statistics, we do not make that emotional connection as readily. We do not feel the reality of what the numbers are telling us.

Nobel laureate, Albert Szent-Gyorgyi, has observed, "I am deeply moved if I see one man suffering and would risk my life for him. Then I talk impersonally about the possible pulverization of our big cities, with 100 million dead. I am unable to multiply one man's suffering by 100 million." Why do you think we are so unable to do this?

We have two ways of thinking about this information. One is a kind of fast, intuitive, gut response to the information. The other is a more thoughtful, deliberative, reasoned analysis of what is going on. Most of the time we are content to rely on the fast, intuitive response, evolved through a long period of human evolution, to help us to deal with what is right in front of us, to protect ourselves and our families and other people around us that we know and love and care about. It does not scale up well in response to many things happening at a distance. It is not that we cannot understand something of the reality of the big problem; it is that we usually do not try hard enough. We do not stop and reflect and do the math, to multiply up from how much we care about one person to how much we should care about millions. We can sort of do it if we work at it, but we usually do not take the time and trouble to make that effort.

It seems like climate change is particularly troublesome in this regard. This fast, intuitive thinking really has very little whatsoever to do with problems that are going to begin to show up several generations from now, in other parts of the world, perhaps some of the biggest problems being for species that we have not even studied yet, ecosystems whose inner workings we barely understand. How are these two different kinds of thinking likely to play out in how we approach climate change?

We already can look at how we behave, and have behaved in past weather crises. Usually what we see is that if there is an extreme event, a severe hurricane or tornado or earthquake, there is an immediate, strong response and a flurry of action to try not only to help the people who have been injured but also to rebuild an area, maybe to strengthen it to prevent future disasters. And that quickly falls away. This is based on the experience of the event and the emotions that it creates. But then when we start to think about the longer-term and the protection of a region against events in the long-term, the response is very weak and falls off quickly after the previous event. So, again, we rely too much on our feelings based on experience, as opposed to the emotions that could be generated by thinking carefully about the problem and the need to deal with it.

And everything you are saying in this interview is based on decades of research.

Yes, I have been studying decision making and situations of risk for more than 50 years, and we have learned a lot. There are a number of people who study this area, and we have learned, fortunately, over a half-century of research. We started off in early research with a concept of risk perception, risk decision-making, as being based on this careful analytic reasoning, following a model put forth by economists. The idea was that we pay attention to the probability of something going wrong and the magnitude of the consequences, and we might multiply these together to get an expected value of the losses. And then we behave in a way to try to minimize this expected loss or to maximize expected gain, a very calculating strategy, a kind of mental cost-benefit analysis. Over the years we came to realize, through various lines of research, that there is more to the story. Most of the time, we do these calculations almost non-consciously, in fractions of a second, through our feeling system, which is evolutionarily honed to be very sensitive to processing information quickly to determine something is good or bad, whether we should approach that animal in the bushes or run away. We have to make split-second decisions. And the brain takes in the information from our eyes or ears, or whatever, and translates it into chemical responses in neurons, that then create feelings. And these feelings are the drivers of action, positive or negative.

Today, we now appreciate the concept of not only risk as something we analyze, but risk as something we feel through our gut. This recognition has come slowly over the years. But now there are a lot of people who work on the feeling system and what is called *dual-process models of cognition*, and that just means two ways of thinking, what Daniel Kahneman, the Nobel prize-winner, has characterized as fast and slow thinking. Most of the time we are on the fast system, because it is easy, it is natural, and it seems right. And most of the time, for what we do in our daily life, it is pretty good. It gets us where we want to go. We navigate our daily decisions fairly well. And we do not have to go through the agony of trying to calculate expected value or expected utility. We do it in the feeling system, but in a very, fast, automatic way.

That is very different from what a scientist would do. They know the value of slow thinking that uses data and statistics, mathematics, formulas, careful calculations. That is the scientific way, though that too is guided to a certain extent by feelings. So, we have these two ways of thinking: the fast, gut feeling, and the slow, careful analysis. But most of the time we are on the fast system. Most of us in the public are operating on the fast system. And that fast system is remarkably effective and efficient for things that we have had direct experience with, or that are right in front of us that we have to deal with. And it breaks down in important ways when there is a distance from the problem, a distance in time or geography or a kind of a diffuseness to the consequences. All the things that climate change entails are difficult for the fast system to grasp.

In what ways might our feelings get mixed up with our objective perceptions of the data on global warming, and how might this effect the way we understand the issue?

Our perceptions of global warming, just like our perceptions of other types of hazards, are driven by images, images in a very broad sense, that includes words as well as visual images. And these are conditioned in us through either direct experience with climate or through reading reports or narratives. The understanding and concern about climate really has to be linked to the ability to imagine what is being described in scientific terms. We can experience things through our imagination, and that is a very powerful way to experience. You can watch a movie about a topic, and you get drawn in, and you respond as though it was real. You get informed and react, and you act in a way as though you had really experienced it. That is a safer way to experience some of the world's phenomena. Experience is critical, and we have to link the information that we are getting from the science. It has to become experiential in some way.

There is a debate within climate circles that is often played out between outrageous environmental activists and careful climate scientists. The careful climate scientists want to protect the authority of science by making very hedged-in, probabilistic statements that, as you have mentioned, are very difficult to feel and do anything about. On the other hand, activists understand it takes something much closer to our daily lives and much more poignant to make us feel about the issue. But they tend to use hyperbole and make claims that cannot be justified by the science, because the science has to be couched in very probabilistic terms, based on the complexity, even the chaotic nature of the biospheric, atmospheric, and oceanic systems involved in climate change. How might we negotiate this process of making climate change real in a very concrete way, without veering from the science?

As you know, from looking at some of the research, we do not do a very good job of thinking probabilistically. It is a way of understanding the world that is fairly recent in the evolution of human thinking. And there are a small number of people, who are very skilled at it, and it is a very important way to make decisions in the face of risk and uncertainty. But it does not come natural. If we are not schooled in the analytic way of thinking about uncertainty and probability, then we resort to fast thinking about certainty and probability. And fast thinking is characterized by serious biases and mistakes in our intuitive, statistical, probabilistic thinking, which have been studied and categorized over the last 30 or 40 years. So, for example, a statistician understands that the accuracy of some statistical value is increased if the sample on which it is based is larger in number. But we often act on the basis of small amounts of evidence; we act as though that it is as certain as large amounts of evidence. We do not operate in the same way a statistician would. There are a lot of examples of that, but in general, we do not handle probability well. We do not handle the concept of a low probability/high consequence event very well. If we can imagine the consequences and it scares us, then we act as though it is much more likely than it is.

So, for example, because we are acting on our feelings, a scary outcome creates in us a sense of anxiety, and that sense of anxiety is not modulated by likelihood. All we know is we feel anxious. And so we confuse the fact that we feel anxious because it is a scary outcome, with feeling anxious because it is a likely outcome. An example of this is in the domain of terrorism, where many types of terrorist acts are so abhorrent to us, and create such strong negative emotions in us, that we act as though all sorts of threats of such things are likely. We throw a lot of money at very remote, low probability, terrorist possibilities, like making very old people go through the same screening in the airport as people who might be more likely to cause harm. We have thrown $1 trillion at terrorism since 9-11 in trying to reduce it in all forms, in all places. Some of it is necessary, but one could say that we have over-reacted to remote terrorist risks. Again, when we think about the outcome, it creates this strong feeling that confuses us about the likelihood. This has been called *probability neglect*, which is an area, where the low probability is treated as though it was higher, because we do feel the consequence so vividly.

In other cases, where we do not have that threat of the consequence, or where we have some motive for wanting it not to be real, we sweep that low probability to zero, and we act as though it really is impossible. Some of our reactions to certain types of climate risk have that character. After the big flood, we sort of say, "Okay, we have had our big flood, now we can go back to normal." And we behave in a way as though there really was not any more risk there.

We have a lot of conservatives today that complain about the regulatory state, but your research suggests that the human animal is so built that it is going to be very, very difficult to respond rationally to these risks, and we need some better guide to our behaviors than what comes natural. Somehow our behaviors need to be guided into a more rational response towards this risk. Can you talk a bit about the opposition to the regulatory state, and what your research implies about how we might want to approach the regulation of risk?

Your question touches upon the issues that Professor Dan Kahan at Yale has studied in the domain of what is called *worldviews*. He shows that all of us have views about what kind of society we would prefer to live in, and if you ask people questions, you can uncover their worldviews. He finds four predominant views that people hold, not exclusively one way or another. In fact, we hold a mix of views, but you can categorize people. One type of person is the *hierarchist*, who would prefer to see society organized in a vertical way from bottom to top, with leaders and wealthy people on top, and others climbing the ladder of success and power. The opposite view is held by someone we would characterize as an *egalitarian*, someone who would like to see the wealth and power of society as broadly distributed as possible. And again you can get at these views by asking people whether they agree or disagree with certain types of questions. Then there are two other views that are also prevalent in the population. One is called *individualism*, and that seems to be what you were touching upon. An individualist believes that people should be free to pursue their goals and ambitions and dreams with as little interference from government as possible. As long as what they are doing is legal, leave people alone and get out of their lives and society will be better off. The opposite view is what we call *communitarianism*. People think that the welfare of a community is more important than the welfare of any individual within the community.

What is disturbing is that when people hold one of these views strongly they tend to hang out with other people who share their views, and they socially reinforce each other. They join together to denigrate any information or people that are threatening their view. They find reasons either not to pay attention to it or to claim that it is invalid. Kahan has studied this in a number of hot button domains, such as abortion and gun control, and more recently about climate change and nano-technology as well. Your individualist and hierarchist tend to pooh-pooh the science of climate change or to find fault with it, whereas your egalitarians and communitarians tend to be more responsive. One of the disturbing things that comes out of this is that when you give people scientific information they then interpret it in a way that is consistent with their view. If it is inconsistent with their view, they will find a way not to change their mind. If you change your mind about an issue like this, it effects your social relations as well, since you have to then justify to your friends, who all share your view, why you no longer think like them. We have had some climate skeptics who suddenly became believers, and they had trouble dealing with their community of friends and followers. It is a socially reinforced type of belief as well as an intellectual one. This is very troubling. And it relates to your point about individualism and ideology, because to deal with climate effectively we have to look out for the welfare of not only our community, but the world.

You have quoted H.G. Wells as saying, "Statistical thinking will one day be as important for good citizenship as the ability to read and write." This seems particularly relevant to trying to think through the implications of our actions on future generations. I am wondering what you think he means here by statistical thinking, and whether you believe he is right about its importance to good citizenship.

Statistical thinking represents the slow thinking that treats information, evidence, and data seriously and uses it to guide our decision making. This is how we have dealt with disease; the great reduction in the prevalence of many diseases has come about through research, through the data, which via statistics, has led to certain actions that have mitigated or eradicated many forms of disease. I think that is what is behind this statement. What we are finding is that all too often we are content with a simplistic form of statistical thinking, this fast, approximate thinking, which makes mistakes in many cases, and leads us astray. It is important, but it needs to be done carefully. It is also interesting that we tend not to teach statistics very much in our schools. I do not know exactly what is going on in curricula these days, but when I went to school I never encountered probability theory until I was an advanced undergraduate. I was taught calculus, which I have used very little, and not probability theory and that sort of statistical thinking, which I think is useful for every citizen in trying to understand evidence about their health, their income, their investments and other sorts of things. I think it needs to be emphasized more and taught from very early ages.

What are cognitive biases, and how might they affect the way we think about global warming?

The word *bias* here is central to the concept. A bias usually means that there is some reaction or some measurement that is not accurate. That implies that you know what the right value is so that you can then see that what is happening is actually a bias and not just another version of the truth. A lot of the early research in the psychology of risk had to do with, as we talked about earlier, intuitive statistical thinking. What was found was that when it is fast thinking, untrained thinking, this intuitive statistics is biased in the sense that it leads people to interpret evidence in ways that a statistician or a scientist would not. There are names for these biases: the *representativeness bias*, the *availability bias*, and so forth. For example, the availability bias means that when we are making a judgment about how likely something is to occur, we base that on how easy it is to imagine the event occurring or how easy it is to recall past instances of that event occurring. What it means in risk is that things that are dramatic and scary are possibly easier to imagine and they create stronger feelings. We go by our feelings, based on how dramatic or imaginable something is, as opposed to what the data says.

And that seems particularly important for looking at our responses to climate change, in the sense that if we have no available examples of how bad it is going to be, we are going to be biased against perceiving it as a very real threat.

Yes, and one thing we have not covered in the discussion is the concept of a low-probability/high-consequence event. If we look to any one spot, with regard to any particular manifestation of climate change, certain things are low probability, but if you look at the earth as a whole, these low-probability events are happening already, in some places, with regard to some climatic events, like islands that are suddenly uninhabitable because they are under water. These damages not only are not low-probability, they are happening. They are certain to happen somewhere, someplace. They are happening now, it is not just in the future. We can become vividly aware of these events, through television and other media, and begin to experience them as they are happening somewhere, to some people today. Maybe that will help to break through this barrier in which we just have to turn away, because it seems unreal.

Setting aside the devastating impacts of hurricanes, floods, or tornadoes, to both property and person, and turning instead to our feelings in anticipation of and following such events, how might living in a world in which natural disasters are far more frequent effect our overall sense of well-being?

That is an important question that I do not think we have a clear answer to. But I do know that we are very able to defend ourselves against unpleasant realities. There are low-probability/high-consequence, and very dreadful things, happening all the time, not necessarily from climate, but for example, from gun violence in our country. We are able to live with that if we can partition it off. For example, there were 513 gun homicides last year in Chicago, but people in Chicago go about their business. Most people are not thinking or worrying about it; they live lives very similar to people in cities that are not so violent. And I think it is because they can partition it off. "Oh, that happens over there in that neighborhood. I am not going over there. I do not live over there. I do not work over there.

I do not have to worry about it." And they do not worry about it. As long as these increasingly frequent events are still not frequent right in our neighborhood, right in front of us, people will be able, for their well-being, to distance themselves. We are very good at distancing ourselves from trouble if we have a sense that we can somehow control our exposure in some way.

Even smokers, who are engaging in high-risk behavior, are able to say, "I smoke light cigarettes, or I only smoke a little bit, and I can control this, and therefore I am not at risk." We are very good at fooling ourselves so we remain comfortable in a world in which things are not so good. It is like the bumper sticker that says, "If you are not depressed, you are not paying attention." So, we do not pay attention. And it will take a pretty intense level of horrible disasters before, I think, most people will significantly be effected mentally or behaviorally.

How is a more vivid sense of the realness of human-induced climate change likely to impact the way ordinary people view their future prospects in life, and how might a constant focus on the issue impact the way environmental advocates and climate scientists think about their own futures?

I think that vivid awareness will lead us to be accepting of the reality of it and receptive to large-scale technological or regulatory actions that could make a difference, even if it is costly or unpleasant. That is why it is important to keep working on the public awareness and sense of reality. One domain that is worth thinking about is the role of literature or narrative or story in helping us see the reality of these phenomena. Narrative is very powerful in helping us to empathize with the victims of various harmful occurrences. Through putting us in their shoes, and understanding their thoughts and their experiences, we can vicariously experience these things.

Some of the surveys being done by Anthony Leiserowitz at Yale show that there already is a level of awareness and acceptance of climate change as a concept. But to translate that into behavioral action, we have to make it easy for people who now accept that reality to behave in a way that is better for the environment. Government and industry have to team up to produce the technologies or regulations that either make it desirable for us to do the right thing or costly for us to do the wrong thing. That has to be coupled with this awareness. You need this kind of multi-faceted combination in order to get any significant societal behavioral change.

ANDREW REVKIN

Andrew Revkin has written several books and has been researching climate change for over two decades. His *Global Warming: Understanding the Forecast* was one of the first books written on anthropogenic climate change. And *The Burning Season: the Murder of Chico Mendes and the Fight for the Amazon Rain Forest* was turned into an award-winning HBO film. He is perhaps best known for his New York Times, Dot Earth blog, which was recently ranked by Time Magazine as one of the 25 best blogs. And he is the only person ever to have been twice granted the Communication Award by the National Academy of Sciences, arguably the most prestigious award in science journalism.

His Dot Earth blog centers around the question of how 9 billion people might live together sustainably. Revkin has relentlessly raised the question of how we might mitigate and live with climate change. And his blog has stimulated a vigorous debate among environmentalists over the difficult trade offs involved in feeding and powering the planet.

He has been described by the Columbia Journalism Review as "incredibly successful at encouraging copious, high quality commenting and debate." Then again, his posts on population control in 2009 prompted Rush Limbaugh to suggest he kill himself to save the planet. Fortunately, he is still with us. Revkin is a multi-instrumentalist and songwriter, who often plays backup with Pete Seeger. He teaches at Pace University.

THEO HORESH: *What sorts of debates is global warming stirring up among environmentalists and which do you believe to be the most important?*

ANDREW REVKIN: Among environmentalists, as opposed to the world more generally, there is a lot of agreement. But there are definitely some big areas of division. The biggest I have seen is a persistent perception among probably the brunt of the environmental community that global warming and the build-up of greenhouse gases is a sort of twentieth-century style problem, similar to acid rain or smog, where the approach that works is you lobby, you litigate, you pass laws, and then regulate - and the problem goes away.

Then there are those, who are a fairly significant part of the discourse now, who see it as something much more profound, which is essentially this disconnect between our energy norms and climate realities, that is not going to be solved through those traditional processes and certainly not through those alone; who see that you have to have a big focus on innovation; that it is going to be impossible to make existing sources of energy expensive enough to prompt the change that would be required if you really hewed to the science.

There are problems with both of those stances. So, like most things, it ends up being sort of an all-of-the-above situation. But in discourse, you often see people resorting to fairly simplified approaches.

The recent debates over nuclear energy and hydraulic fracking remind me of the old joke about the way to conduct a leftist firing squad: stand in a circle and shoot. How do you explain the fierceness of the debate we are now witnessing among some environmentalists, and how do you think we might steer our way through some of these debates with wisdom?

Another reality of twentieth century environmentalism was there had to be a bad guy, and there are bad guys – I have written about quite a few. There are companies and interests who are dead set on maintaining fossil fuel norms, and who are not being fully honest in how they make that happen, whether they are doing it through lobbying or disinformation. But again with climate change, you have to look in the full circle at who is driving the demand for fossil fuels. I live in an old house in the Northeast, we have oil heat, I looked at alternatives, they were really expensive. Geothermal and solar panels are still expensive, even with the subsidies that are available, and I drive a Prius, but we also have a minivan. I would love to have everything be neat and clean, but it is not possible, so I go to the gas station every few days. Is Exxon the only villain? Not really.

But it is hard to acknowledge those realities sometimes if you state your goals in pretty hard and fast ways, that it is all Exxon and duplicity and big companies. Frankly, you see the same dynamic in the food arena with GMOs. It is mostly people's antipathy to Monsanto that drives their stance on things like GMO technology, as opposed to the actual science. And in areas where there is scientific uncertainty or a lack of information, like with fracking - in areas where there is a paucity of data - assertions essentially dominate. Climate change is another one of these issues, where the things we know profoundly are very clear, but the things that really matter are not. Things like how fast sea levels are going to rise through this century are still very uncertain, as the new IPCC report makes clear. So, that leads to all kinds of ugliness in the debate.

It seems that when environmentalists recognize the extent to which they are a part of the problem, through the various elements of their lifestyle, whether through eating meat or driving or using energy, many seek to resolve the tensions by becoming rigid ethicists, trying to clean up the messiness and reduce their carbon footprints to the min-

imum possible degree. Then you see others that kind of accept themselves as part of this system, coming out with a more lukewarm sort of reformism, or they go into clean energy and make a career out of it. What I do not tend to see among those who acknowledge the extent to which they are part of the problem is the same kind of vigorous, you could say radical, push for doing something about climate change and other environmental issues as you get amongst those who have a bad guy. Can you talk a little bit about the tradeoff: what happens when we lose the bad guy? How can we have a vigorous environmentalism without always having to attack someone else?

That is a huge challenge, and I have written about this very question in the context of nuclear power, where I was moderating a debate over this recent film, "Pandora's Promise," which points out the upside of nuclear energy. During the panel I said, "It is hard to build a movement around the phrase '*some* nukes.'" It is really easier to say "no nukes"; and it is kind of easy to be just pro-nuke. But to say, "Some existing nuclear power plants are mostly flawed, but we need a push to have new ones, and we have to deal with the waste issue" - there is no simple slogan. How do you go marching for "some nukes"? And you can look at other parts of the energy mix and have the same outcome.

Then if your focus is on American energy policy, which a lot of environmental groups still focus on, how do you acknowledge the reality that over 90 percent of growth in emissions of greenhouse gases in the next several decades is coming in India and China and other fast-emerging, developing countries. It is sort of facile to say, "If we do it, then they will follow." There is no evidence of that actually. Given their political realities and the energy poverty in big chunks of their populations, the imperative will be to supply energy before the climate imperative kicks in. And that makes it really hard, too. I am not saying give up, and I am not saying reject those who are purely focusing on American policy. But the reality is that it is super "wicked," in the most formal sense of the term "wicked problems." This is beyond wicked, in a special category.

It seems there can be this tacking back and forth between solutions among environmentalists. We recognize that this is a global problem, so we want a global solution. But we recognize that a global solution is too hard to achieve – it is just too hard to get everybody on the same page, particularly with such divergences of interests and levels of economic and democratic and social development. So, then we go back to these small-scale solutions. But the small-scale solutions do not always scale up, and they often do not account for the fact that they might not have any influence globally.

Having done this for a really long time, 25 years or so, you have probably seen a lot of bouncing back-and-forth; I am guessing you have done a lot of bouncing back-and-forth regarding your own best solutions. When you look at the wickedness of the problem, and the difficulty of scaling-up the local solutions, the challenges of getting a global solution, where does this drive your thinking?

I have been liberated to some extent on this in the last five or so years in shifting from thinking of this as a problem, meaning a problem that needs a solution, to being something more akin to public health or poverty - it is an issue. The human relationship with climate and the idea that it is a two-way street is in the early stages of kicking in and becoming a reality for most societies and most individuals. It is much more akin to something that should be addressed essentially as part of how you look at any choice, whether it is where you live, what you learn, or what you do with your career. If you care about poverty, you tend to focus your career on things that would lead you to reduce poverty. You do not have a sense, though, of a specific outcome, other than year-to-year keeping a sustained focus on the issue.

So, the more we shift to a sort of normalization, that climate change and climate vulnerability are really important things to consider in making any decision - whether it is at the local level of planning or how you retrofit a house or where you live or who you vote for – the more they become simply factors that we weigh going forward. The way to have a sustainable relationship with the climate and energy is to consider it a journey that has implicit uncertainty and wickedness. In a case like that, the best way forward is to focus on maximizing the traits, in society or in yourself or in your own household, that maximize the potential to have better outcomes - but without knowing specifically how it is going to play out. And remember that the knowledge on global warming is still evolving in terms of how warm it is going to get.

Locking into a specific mindset that a specific solution is the thing to pursue is fine for some. And I am all for the silver buckshot approach, as opposed to a silver bullet. It is not just about technology but also about the choices we make as individuals on how to be active, how to pursue a goal. But, personally, I just see it more as an issue that we work on, not something we solve.

So, we do not have to have answers, rigidly adhere to those answers, and then push hard for each of them, because there is a danger that information will change. The political playing-field is going to change over time and we need to keep re-adapting. We want to integrate asking these questions, pushing ourselves in multiple different directions, in our lives, in our society, in our governance. But we are going to have to use a multitude of different solutions, and we do not really know what is going to work at this stage, no?

There are some who would say this is a cop-out. In fact, I have had people rhetorically punish me for not choosing a number - 350, 450, or 550 parts per million of CO2. But given global trajectories, choosing a number is silly if all the arrows are pointing in the opposite direction anyway. I think we have a lot of work to do before we close in on a number. And I think we need to acknowledge that this is also a multi-generational reality that is not going to be "solved."

Especially going into Copenhagen, the climate talks of 2009, there was this seal-the-deal kind of approach, that there was a particular opportunity because of a particular President and a particular round of negotiations, that had to be achieved - that is so over. Hopefully, people have recognized that this is not one person's magic bullet. It is not an Obama problem to solve. It is a changing relationship with energy and with the planet around us. It is going to take time and involve more than one generation.

Now a lot of environmental leaders, and everyday environmentalists, seem to want us to freak out about the issue. They seem to think that in freaking out we are going to accomplish something special. They seem to want us to feel anxious, to move quickly, to act with vigor. And yet there is a lot we can lose in freaking out. For one thing, it is very difficult to sustain over time. For another, if you encourage people to freak out, and you are freaking out yourself, not a lot of people will listen - so you lose some authority in the process too. I am wondering if you can talk about the demeanor we bring to our environmental activism and concern.

I spent the first 20 years of my writing on the greenhouse gas problem, focused on it as a bio-geo-physical problem: gases are going into the atmosphere that trap heat; that has consequences for the climate system, the oceans; the CO2 has chemical impacts, and this is all changing the way the world works in X, Y and Z ways. Then it was around 2006 that I started digging in more on the social sciences as they relate to how we perceive problems, especially ones of this type. And that work was way scarier than anything I learned about glaciers, or standing on the vibrating sea ice at the North Pole. It is much more daunting when you look into the behavioral sciences and you realize just how poorly attuned we are to getting this problem right.

But it has also made me realize that I cannot begrudge the fact that some people are way more freaked out about it. There is probably a mixture of failure on my part, and a failure on their parts, in terms of who is appropriately freaked out or what is the right level. There is no right level. We are a variegated species with edge-pushers and door-lockers, and some of us are different than others in pretty basic ways, so getting comfortable with that is important too. Yet to my mind, from what I have learned from the literature, this kind of run-to-the-ramparts, ring-the-bells approach, on a problem of this sort, is bound to fail.

And actually, in terms of my reading of the science, I do not think it is that kind of problem. Again, humans are going to persist. We will find a way forward. The data do not point to, despite a lot of rhetoric, imminent tipping points. There are some who will still contend they are there; I do not see that as being in the science. That does not mean that there are not what Steve Pacala at Princeton calls "monsters behind the door," unknown unknowns, or hinted unknowns, things we think are possible. But again, my reactions to these scenarios and to the data are different than others and that is just the way it is.

So, what concerns you most about climate change, and how do you weigh these worries against other major global challenges?

The biggest issue with the greenhouse effect, as amplified by humans, is that it is really hard to ratchet back. It is like a one-way knob. The gases that matter the most, carbon dioxide particularly, are very long-lived. So, once you add new carbon that has been stashed in the ground for millions of years, it is not going anywhere for thousands of years - it will be in circulation. Concerning capturing carbon dioxide from the air, there are costly ways to do it, but how you do it affordably, on the scale of gigatons, is really hard to understand. So, that is the big thing to me; it is momentous. As Susan Solomon and David Archer, in his book, *The Long Thaw*, have laid out convincingly, this is not a short-term issue. This is another reason why it has been valuable to move away from the idea it is a problem we are going to solve. We have already nudged the system in ways that will have impacts for thousands of years. We have essentially staved off an eventual Ice Age that was probably coming fairly soon, and that is a big deal. It is the momentousness of it that is remarkable to me.

Recent research shows that different personality styles react to the data on climate change in different ways. Can you say a little bit about this research and why it is significant?

Work by Dan Kahan at Yale, and others, shows very convincingly that more science does not clarify disputes over something like global warming. It actually amplifies them, because we all tend to be either sorting information in ways that fit our predispositions, or we choose our sources based on our tribal affiliation. And tribal affiliation, whether you are a libertarian or liberal or something in-between, is a very powerful force in how we behave. Basically, there has been work that has shown that the positions of people, who are most energized by global warming, are hardened by more data - whether they are skeptical or outright reject the science or are extremely scared about global warming, in a way that is not reflected in the science. If you are not tracking that work, you can end up really bashing your head against the wall trying to fill the public's heads with more information, more science. The term for this is "the deficit model." That just does not work.

So, how do these personality studies affect the way you approach the issue? What do you do with that?

Well, I look for areas of agreement. They are there, and the same research that shows this pattern pretty convincingly also shows that people who might be very divided about global warming can be very much in synch on issues like energy efficiency. That means that if you approach the issue first from a sort of energy-wisdom track, then you can end up with some surprising confluence. For example, the Heritage Foundation would like to end energy subsidies, period. They do not want anything subsidized - wind or fossil fuels. And there are plenty of environmentalists who want to end energy subsidies, at least on fossil-fuels. Many of them still want them for the lagging technologies, like wind and solar. But at least there is overlap there, and to me that says something important. At least you can have a conversation about subsidization of energy in America.

The same thing goes for making communities resilient to climate extremes. Over and over, I have seen pretty convincing arguments from libertarian groups that hate that we are subsidizing federal insurance that keeps people building in harms way, such that they get bailed out by the federal government. And there are a number of environmentalists, or people concerned about climate change, who would say the same thing. And this goes for wildfires - I have written about this as it relates to what happened in Colorado the last couple of years. We have this fast-paced development in areas where there really should be no development. And it is only there, at least partially, because second mortgages are deductible. Why should we have a federal-tax deduction for a second-mortgage to build a house in a wildfire red-zone? That is a policy that needs to end, and I know libertarians who strongly agree. So, there are ways to get at the core challenges, where you see overlap, and I think that is one way forward.

Of course, the reality of our politics is such that it is still driven from the edges, and the Tea Party is the ultimate example of that. Nothing I have said implies that the Tea Party types are going to say, "Okay, we can agree on subsidies." Because it is all about strategic advantage and legislation - and the same for Liberals. For many people in the Beltway, this is ultimately a game of *Stratego*, with the outcome being influencing Congress in one way or another. That is an arena where the advantage is all with those who hold the high ground, the *stastists*, as I call them, people who want to maintain the status quo on energy. Basically, society is very fat and happy on fossil fuels. Anyone who wants to change that norm has a huge task, and those who want to sustain it have a pretty easy task. It is a completely asymmetrical challenge.

Have you found on your blog and in your personal conversations that your interest in highlighting overlap between Left and Right, between climate change activists and climate skeptics, has yielded much response from the climate skeptics and people on the Right?

The reality, to me, is that there is no such thing as climate skeptics as a group. There is a range of people who hold a whole array of views. I really butted up against this a few years ago when I covered one of the Heartland Institute Climate conferences. They were having all these arguments among themselves about the science. They all had different theories for why the science is wrong, and they were having these fights, just like you would see at a science conference. But the one thing that brought them together was their antipathy to regulatory solutions to environmental problems. They are all basically limited-government people.

The people who reject or challenge the seriousness of global warming fall into a range. I think they mostly have a libertarian substrate to their skepticism on the science, and then some of them are just there to rattle the same talking points that have been discredited for a long time. They are pretty serious and engaged and smart people. They do not reject the science so much as the interpretation of it as pointing to calamitous outcomes. So, it is hard for me to even address the question about climate skeptics without starting with the reality that there is a range. And this goes for the climate concerned as well.

Now in 2004, just after Bush got elected for the second time, I set out to write a book called, "The Healthy Conservative." I got deep into writing the book, but ultimately gave up, frustrated by my lack of examples of what I considered healthy conservatives. As a progressive, I was trying to reach across the aisle and look for something healthy and good and of great integrity in my political opponents. And one of the things that happened to me in the process was my circle of moral concern diminished. In the process of trying to reach across the aisle and speak the same language, I had to cut out a number of environmental concerns; concerns for species loss, concerns for non-human life, concerns for distant generations. I found myself beginning to talk more in terms of national well-being. And I would even use progressive arguments for bringing about greater national well-being. But I stopped talking about poor people in the Sahil of Africa and about distant generations, and this changed the way I thought and felt.

And I see this happening with people who often appear to be more reasonable environmentalists, who are willing to reach across the aisle, to have dialogues with this spectrum of people who have some skepticism about this or that aspect of climate change or climate change activism. And I am wondering what you have observed happening to the arguments of the people who are willing to take this softer approach, who are willing to reach across the aisle. Do you see this kind of shift away from concerns with future generations, concerns with the extremely poor, concerns with non-human life, or are they able to sustain that wider ethical circle of concern that has so characterized environmentalists for the last several decades?

I know a lot of conservatives who are very focused, in part through religious affiliations, on the welfare of the poor or especially disadvantaged people in developing countries, who go down and rebuild houses in New Orleans or go overseas to do the same kind of thing. I have seen pretty strong passions of that sort. John Christy is a climate scientist at the University of Alabama Huntsville, a garden-variety climate skeptic on the science, in the sense of questioning the seriousness of global warming as it has been portrayed by the brunt of the scientific community. And it is interesting that he was a missionary in Africa for a chunk of time in his early life, and every time I trade emails with him or talk to him, some of that comes through. His passion about the need to bring energy to poor people, by whatever means - meaning a cleaner lamp or a cleaner stove, even if it is using a fossil fuel - is driven by his ethics.

So, that is where you get this sort of disconnect. There are more than a few people that I have run into with that same kind of zeal for human welfare. But their logic is that the energy imperative dominates the climate concerns, and that is the basis of their rejection of aggressive actions to move away from fossil fuels. So, the ethical, even the multi-generational ethical, imperative can lead to different conclusions about climate, depending on who is doing the thinking and feeling.

Now, I do see what you are saying, particularly in regards to the climate debates - that often times the conservatives will come out showing greater concern for the extremely poor in developing countries. I am much less likely to see conservatives express concern for generations far into the future, though. And I almost never see this sort of concern for species loss and particularly the welfare of non-human life. It seems so often that what moderate environmentalists are encouraging radical environmentalists to do is to stop talking about concerns with non-human life.

Again, there is a lot of variability. I know people who care a lot about animal welfare, particularly separately from biodiversity welfare, who are conservative. But the core issue I think you are talking about is acting on something like climate for the sake of biodiversity. There too there is a range. I live in the middle part of the Hudson River Valley, where I have some passionate hunter friends who are definitely not Democrats but who care hugely about sustaining thriving environments so that they can be utilized in a respectful way.

The thing that really drives a lack of concern for biodiversity is inequality more than anything else. The bush-meat trade is driven by the rising middle-class in African cities, who still want to eat monkey meat. But part of it is logging crews, who just want to eat something reasonably good for dinner. So, you end up with this empty-forest syndrome, where they are killing off the wildlife even before the trees have gone down. And the real driver of biodiversity loss in many instances is poverty.

So, as you are disagreeing with me, I have a smile on my face, because I am appreciating the way you are widening my own perspective. But I am well aware that when many other environmentalists disagree with you, they are pretty stressed and angry. I see you sitting in the middle of this cauldron of debate amongst environmentalists, some of the most fierce debates around fracking and nuclear energy and the severity of climate change, not only holding a space for it, a very valuable space of debate and dialogue, but also sitting in the center of all of that with a rather cool and relaxed demeanor. Tell me about your facilitation of this dialogue amongst environmentalists and how you are able to maintain the calm in the center of the storm.

Oh God, I do try once in awhile to unplug. I grew up as a middle child, so I have a middle child's view of the universe - facilitator, mediator. And life is simply too short to get too angry. We have to work on these things together.

The ultimate example of this came this summer when I was on a panel with Josh Fox, about his new film, *Gasland 2*, with Alec Baldwin, the actor, who is very anti-fracking. And Josh Fox, of course, is very anti-fracking. And we were there with one other journalist, who was also anti-fracking. And I had a microphone that was working horribly, so things were really bad. But in the end Josh pulled out his banjo, I pulled out my guitar, and we played a coal mining song together. And we smiled and had dinner afterwards. And there was some tension, but…

There are people out there who are fundamentally working to destroy the planet. And there are actually a small number of people, who are in the camp of being completely dishonest, malicious, Dr. Evil, planet-killing folks. But the vast majority of people I have found have some sense of wanting to make the world a better place, or at least not as bad a place. And you can find common ground. And whether it is pulling out a banjo or guitar, or finding a way to look at some issue where you have a shared view, you can find a way forward.

I think a lot of these little shifts in trajectory can add up to big things over time. So, it is all about sustained attention to better outcomes and focusing on the reality that we are on a journey, that this is not a problem to solve, but a reality to attend to. That is my approach - and I am trying not to say it is the right approach for anybody else.

GEORGE LAKOFF

A founder of the fields of Cognitive Science and Cognitive Linguistics, George Lakoff is the best selling author and co-author of eleven books, including *Don't Think of an Elephant* and *Metaphors We Live By*. He has published hundreds of scholarly journal articles and has applied his ground-breaking research to a wide range of fields, including linguistics, politics, psychology, poetics, mathematics, and philosophy. He spent over a decade as Senior Fellow at the now-folded Rockridge Institute, which he founded in 1997, and whose stated goal was to strengthen democracy by providing intellectual support to the progressive community.

Lakoff is best known for his work in developing the concept of linguistic frames and for the application of that work to political discourse. According to Lakoff, "frames are the mental structures that allow human beings to understand reality - and sometimes to create what we take to be reality." These frames tend to come in the form of metaphors, and the metaphors through which we experience the world have moral implications. Hence, metaphors like "up," to signify better, and "down," to represent worse, are value-laden and integral to human thinking. These metaphors are laced throughout every discussion of global warming.

Professor Lakoff applies his linguistic expertise to help progressive citizens' groups, activists, and policy makers think through their values and general principles and to strategically formulate their policies. He has advised several Democratic Party officeholders and candidates and has spoken at the caucuses and retreats of both Senate and House Democrats. He currently teaches at the University of California at Berkeley.

THEO HORESH: *Al Gore has his "head in the clouds." Climate science is "hazy and ungrounded." The results of studies on climate change are still "up in the air." The nations of the world need to "sit down together" to forge a climate deal. Tell me about how these sorts of metaphors that are embedded in climate discourse influence the way we think about climate change.*

GEORGE LAKOFF: Let's start with "climate deal." The metaphor of a deal assumes that climate is an economic matter and that it involves give and take on the money involved - and that is a very questionable assumption. This has to do with the future of the Earth, not just about short-term negotiations of the particular people who happen to be in a negotiation and making a deal. What that metaphor does is frame the discussion in economic terms. There are many cases where the issue of global warming is framed in economic terms, and that can be appropriate sometimes, but it is often inappropriate. Then there is the idea that climate scientists have not done their jobs, which they have. There is no doubt that global warming is increasing. The questions that need to be asked are why people say this, why there is so much resistance to it, mainly from conservatives, and why they are so successful at making it work. At the same time you need to ask the opposite question of why the framing of environmentalists is doing so badly. Those are complicated questions.

What sorts of frames do we tend to use when we think about global warming?

There is the question of how to "transition." The assumption is we have to have an intermediate transition to real sustainable energy, something like natural gas, which is not as bad a fossil-fuel as gasoline. But it is still a fossil-fuel, and it is gotten out of the ground by fracking, which is a disaster. The fracking is never mentioned and the fossil-fuel is never mentioned. It is just that natural gas is seen as clean. And you cannot say natural gas is clean, it is just less dirty. But it is dirtier if you consider that you get it out of the ground through fracking. The idea of natural gas as a dirty fuel, just somewhat less dirty, is not mentioned and it needs to be mentioned.

Then we tend to talk about things like "climate change." The idea of climate change was introduced by George W. Bush. His framers said, "Global warming sounds dangerous and sounds like people are doing it. Climate sounds nice, like palm trees, and as if things just happen." So, climate change got substituted and progressives started using it, which was not a great idea. Climate scientists have done wonderful scientific work but have hurt the environmental movement for a lot of reasons. The major reason is they do not understand how people understand causation. One of the first things you find out when you start studying causation as a linguist is that every language in the world has direct causation in its grammar. So, you can say things like, "I picked up the cup," which is *direct causation.* What you are doing is exerting a force on something to change it right there and then, which is direct. What really happens in the environment is *systemic causation.* That is, you have an environmental system, an ecological system, and causation is not just direct.

When there was a major snowstorm in Washington D.C. a few years ago in the winter, the biggest snowstorm in a very long time, the conservative people in Congress said, "There is no global warming, look at this huge amount of snow." Now if you think about it from the point of view of systemic causation, what happens? Global warming increases the amount of moisture over the Pacific Ocean. That moisture blows by winds towards the Northeast. It goes over the pole and then in the winter comes down over the U.S. in the form of snow, more snow than you have ever seen before. That is, big snowstorms can be systemic effects of global warming. But that is not discussed in the media, and it is not discussed by progressives, and it is not discussed by climate scientists.

This became clear to me in 2005 when I was at the Aspen Institute for a meeting on the environment. A very famous climate scientist gave a talk, an excellent talk, and he was asked afterwards whether Hurricane Katrina had been caused by global warming. And he basically said, "Oh no, we cannot possibly say that - it is a probabilistic event. We cannot predict any particular hurricane and we cannot predict exactly where any particular hurricane is going to be formed and hit." That was the wrong answer. The right answer was, "Global warming heated the Gulf of Mexico 3 degrees, which creates a huge amount of energy. The frequency and intensity of hurricanes is a function of the heat energy in the Gulf of Mexico, so a certain number of very powerful, very wet hurricanes are going to happen as a result of global warming, and a certain percentage of those are going to go towards New Orleans." That was the answer that should have been given, that this was a systemic effect, not that you could not tell. The same thing happened with Hurricane Sandy, which was systemically caused by global warming.

Now it seems that systemic causation is more complex and therefore more difficult to think about than direct causation, but it also seems like your use of metaphor is able to simplify this complexity. Can you say a little about the ability to make the complex simple?

It is not merely that it is more complex, it is that our conceptual system is formed when we are children, and the experiences we have of causation are mostly direct causation. That is a very important fact about what goes on in the grammars of the world's languages and how you are taught about the nature of causation. But scientists are also taught causation in that way. Scientists should know better, especially people who study ecology.

You have to learn explicitly about systemic causation. But it can be taught, because systemic causation can be broken down into a couple of parts. First, there is network causation, namely sequences of the sort we just saw. Second, there are feedback loops. For example, the melting of the Arctic ice cap is a feedback phenomenon. And there what happens is the ice cap reflects heat and light. As it melts, because of global warming, it reflects less and less heat and light, giving an exponential effect, which is a feedback effect. And the third is probability. Those are three things that can be taught in the upper levels of grammar school, certainly in high school. There is no reason why every scientist cannot talk about systemic causation, but it does have to be taught.

Why do so many people believe fossil-fuel emissions are not causing temperatures to rise when even many of the climate skeptics dispute only the seriousness of rising temperatures?

There is an old saying, "You cannot do anything about the weather." Namely, there is an assumption that there are things called *natural causes* and that the weather is one of them. You learn that weather is a force of nature. And it is very hard to understand that human beings can collectively cause things, especially since direct causation is about individual human beings. It is hard to think about what collective action by human beings can do, for two reasons. One, there is the question of systemic causation. And two, the fact that when you learn direct causation, you basically learn it about individuals or maybe small groups, but not about all human beings on the planet - you do not learn that. And this is, again, something that has to be taught, because it goes against the most basic understanding of causation that we learn automatically growing up.

This view of causation is *physically* in our brains. Every idea you have is physical. It is there in brain circuitry. And you have to learn when your brain circuitry is failing you and when you need other concepts. And that is not easy to teach, and it is not easy to notice.

There seem to be numerous spillover effects of teaching systemic causation in a complex world, because it is not only a problem of people not understanding climate change, or any number of major environmental challenges, but rather the world itself. In other words, teaching systemic causation would help us to understand the world itself and our place within it because there are massive systems influencing any given event in the world.

That is exactly right, and there is another problem that you have on the environmentalist side. Numbers do not mean anything unless they are framed in a way that can be comprehended, and most of the talk about numbers is not, so if you really learn about these things, and people who are very serious about the environment do, then you will say, "Oh no, we hit 400 parts per million, and that is a terrible thing." But the fact the New York Times says we have hit 400 parts per million means absolutely nothing to most people. The numbers need to be framed in a way that people can comprehend. They need to be made comprehensible, and that is not easy, because the whole idea is systemic. There is another problem and that is the word *environment*. The word environment means something outside of us, but nature is inside of us. We breathe air, we drink water, we eat food, we ingest things. And not only that, our connection to nature is inside of us in ways most people do not even understand because of work in neuro-science that most people have not heard of yet.

Part of the research in what are called *mirror-neuron systems* has to do with what are called *canonical neuron systems*, which are located a few millimeters away from the mirror neurons in the pre-motor cortex. A canonical action is a normal action you would take, like pealing a banana as opposed to sticking it in your ear. What was discovered was that the same neurons function when you see an object as when you perform a canonical action on it. You see a banana and the same neurons are active as when you would peal it. You see a glass of water and the same neurons are active as if you would drink it, but not as if you would pour it over your head.

That is important because we evolved to connect with nature and to have automatic canonical actions that we performed as we went around the world. It is still there, and it is a link inside of us to the external world. And most people do not even know that they have in their bodies something that links them to the natural world, that nature is as much inside of us as outside of us.

It is difficult to picture the science of global warming. We cannot see the whole world. We can see static pictures of it, but we cannot see change in the whole world. Nor can we see if our own local climates are changing over time without the use of statistics. So, how can we simplify the science?

Every weather forecaster in the world should be talking about these things. People who are professional meteorologists and people who go into weather forecasting should be trained to think in these terms. That can be done, and there are ways to visualize what is happening. For example, you need to know that this huge tornado that happened in Oklahoma had more energy in it than almost any tornado in modern history. And the amount of energy you get in a tornado correlates with global warming. And the correlation between global warming, tornados, and droughts in the Midwest is astonishing. There are huge correlations that you can show on a graph. What is the heat now versus before, what are the droughts now compared with before. Here is your heat and drought graph, put constantly on every weather forecast. That can be done, newspapers can post pictures, it can be put all over websites.

There is no question that we can get these ideas across. But it takes understanding how people learn and how their ideas change. They do not change by direct confrontation with our deepest held ideas. They change by going around them and setting up different neural circuitry. You have to know how to do that, and you also have to know that the facts in themselves do not necessarily help.

Back in the seventeenth century, the Enlightenment said, "Everyone is human, we are rational animals, we all have the same reason, and therefore, we do not have to let kings and Popes reason for us - we can reason for ourselves." That was very important in the development of democracy. However, that view was technically not necessarily right. Descartes assumed that reason could not be physical, because if it was, then you would not have total free will. It turns out we do not have total free will. Because of certain strong framings in our brains, the facts will not be heard, nor made sense of. Enlightenment Reason says you just have to tell people the facts, and they will all reason the same way, via logic, and get the right answer. That keeps not happening, over and over. And yet, progressives who believe in Enlightenment Reason, and who are trained in it, and who are trained to do policy studies using it, keep making the same mistakes, over and over and over.

Conservatives tend not to make those mistakes, and they do not make those mistakes for an interesting reason: many conservatives go to business schools, and in business schools, they teach marketing. Marketing professors study how people really think. They think in terms of frames and metaphors and images and narratives. However, if you go to departments of political science, public policy, economics, and law, what you find is that people teach Enlightenment Reason, the rational actor model and so on, which is not the way people really think.

What you have on the progressive side are lots of very smart, very decent people, who believe that all you have you do is tell people the facts and to reason logically, and they will get to the right conclusion - and that is simply not true. You see this especially on progressive television shows. You turn on a progressive television show and the host will say, "This conservative commentator said this, and here are the facts that show it is wrong." But when you negate anything, then you activate it. If you say, "I am against the idea that global warming science is no good," then it brings up the idea that global warming science is no good. The idea that you can just go out and contradict what a conservative says and this will convince people is simply false, because when you do that, you help conservatives by activating their ideas.

It seems that as each academic discipline develops, each generation sees more doctorates, and our knowledge of the world increases, we are getting more and more counter-intuitive answers about the way things really work. What is the role of Enlightenment Reason in rationally thinking through these counter-intuitive results?

You need a new enlightenment that tells you what real reason is about. There is something called the *enlightenment reason bias* - you have to watch out for it. You see it in Daniel Kahnemann's work on behavioral economics, where he claims there is *thinking fast and slow*. And by thinking fast he means this unconscious, automatic thought that biases you in certain directions and leads to what he calls *cognitive biases and irrationalities*. But then there is rational thought that is slow and deliberative and uses the tools of reason, like the rational actor model and logic and statistics and probability theory, and you should be using those.

There are two major mistakes here. First, the unconscious reasoning that is really there, and he is right that it is really there, uses frames, metaphors, narratives, etc, And you absolutely need it in everyday functioning, because you do not have the opportunity to use other kinds of tools. You have to make judgments, often very quickly, and you have to make judgments without total information. You have to make judgments where someone else has more information than you have, you have to make judgments within your own brain. Now if you are using that system of thought throughout your life and it is working okay, then it may be appropriate to use it. But it may not be appropriate to do that with global warming, and you have to know the difference. So, you have to have some sense of what your own brain is doing unconsciously. What we need to be teaching, among other things, is how unconscious thought works and what unconscious models we learn about the world that are not true. And yet, we are going to learn them no matter what, so you have to watch out for them. And the reason we are going to learn them no matter what is that we are going to learn them as children, and children's experience is not going to teach you about systemic causation.

Now you have developed two metaphorical models of how we in America think politically, the strict father and nurturant parent models. Can you talk a bit about how these function?

The idea is that there are two kinds of moral systems that are based on families. Morality is about well-being, and there are certain experiences that children have of well-being. In general, you are better off if you listen to your parents. And in general, you are better off if your parents nurture you and take care of you. Listening to your parents gives rise to a metaphor that morality is obedient to legitimate authority. The nurturant experience leads to a different metaphor, namely that morality is nurturance. And these give rise to two kinds of families.

In a strict father family the father is the legitimate authority. He is supposed to know right from wrong, to know what is best. And his job is to raise children to listen to him, to be moral beings because he knows what is right and wrong. When they are born, children are not moral, because they have not learned right from wrong. Conservatives talk about the feel-good morality of liberals, as if liberals were babies who have never been trained to be moral by their strict fathers, because they did not have strict fathers. The idea here is that the strict father teaches morality, and he does it by punishing children when they do wrong. The idea is that the punishment must be painful enough so the children will not do wrong and will learn the internal discipline to do what is right, namely what the strict father tells them. Otherwise, they will get punished painfully. If they do that, there is a logic, namely that they will become moral beings from that discipline. And if they are disciplined, they will be able to go out in the world and earn a living on their own. The assumption, therefore, is that if you are poor you are not disciplined enough to earn a living on your own. And if you are not disciplined, you cannot be moral, so you deserve your poverty. That is a kind of argument from strict father morality that you hear all the time from conservatives.

The idea here is that because you are first governed in your family, you learn a metaphor that a governing institution is a family. This does not just apply to government, it applies to schools, it applies to religions, it applies to athletic teams, and it applies to the market, all of which are governing institutions. That is, the market is seen as moral and natural. It is seen as moral because, if you believe the conservative interpretation of Adam Smith, that if everybody pursues their own profit the profit of all will be maximized, then that seems moral. It seems natural because it is assumed that people are naturally greedy. If you put those two things together, the market is a kind of strict father, and it imposes discipline. You have heard the expression, "Let the market decide." The market is the decider who decides who gets rich and who does not get rich. And the market should be that decider because it is moral and natural. What does that mean? It means that just as in a strict father family no one should have authority over the father, so in the market there should be no institution that has authority over the market, like government. So, there should be no regulation, no taxation, no worker rights, no unions, and no tort cases. And those are exactly the things that conservatives are against economically. The strict father interpretation of the family is then projected onto all major governing institutions.

This has a tremendous effect on democracy, because from a conservative point of view democracy gives someone the liberty to pursue their own interest without interference from anyone else and without help from anyone else. Now what does that mean for the environment? It means that if you own property you can, therefore, do anything you want with it. A corporation can go and buy up vast amounts of land or forests or rivers or streams and do anything they want with it in that view of democracy.

Now there is another view of democracy that comes from nurturance. In a nurturant parent family the parents have to empathize with their children. They have to know what they feel and what they need and what they want. They need open two-way communication. And they have to take care of themselves too, for they can not take care of someone else if they cannot take care of themselves. They have to empathize and act on that empathy. They have to be responsible and work as hard as they can on it - that is a nurturant parent family. They have to teach their children to be exactly the same, namely their children have to take care of themselves and also have responsibility for others. That is central to any nurturant parent family.

What does this say about institutions? It says that democracy arose because citizens care about other citizens. And they exercise that care responsibly through their government. What the government does is make public provisions. That is, from the very beginning of our country government was in charge of roads and bridges, a national bank, a patent office, judges, courts, police, and interstate commerce. All of that was there from the beginning and it expanded. It expanded because more things were needed to make life good for people. You had public education from the beginning. You needed public health, you needed research supported by the government, which is why we have computer science and satellite technologies and all of the things that go with them. Without that, you cannot start or run a company. If you do not have access to transportation, to roads and bridges and airports, if you do not have electricity from an electric grid, if you do not have educated workers to hire, you cannot start and run a company. You cannot do anything. You cannot even live a decent life if you do not have support from your government given to you by your fellow citizens, who care about other fellow citizens. Now that goes against the conservative view of liberty that says you can do anything you want with nature, with your land, for it is there to be exploited by you for your profit.

The idea of the nurturant view is that you need to have empathy with the natural world. You need to understand that you are connected to the natural world. And you need to have empathy for all other people, now and in the future. That is a totally different view. That is a progressive moral view and that progressive moral view lies behind environmentalism. Environmentalism is an inherently progressive enterprise. And the morality behind it is often not stated, but it is there.

One of the interesting things about the strict father/nurturant parent distinction is that most people have both views in their heads. Both moral systems that contradict each other are there in neural circuitry. And a type of neural circuitry called neural inhibition says that when one neural circuit is on it turns the other off. Most people have both of these views and they can be affected by language, because every word is defined in language relative to a frame. Language activates frames; therefore, if you choose conservative language you activate conservative frames and those frames activate the moral system. Why do they activate the moral system? Because all politics is moral, because when every political leader says, "Do this," the assumption is because it is right. No one gets up and says, "Do this because it is wrong and does not matter." The assumption is that it is right, but they have different ideas about what is right.

When I first read your Moral Politics, where you lay this out for the first time in the nineties, I could not tell whether you were a liberal, a conservative, or sitting on the fence, because these two models sounded so equal. I imagine them side-by-side, co-existing together in a happy family with a strict father and a nurturant mother. Both systems appear necessary and complimentary.

But there is another way we can lay out the two models. That is, the strict father morality precedes, in a developmental sequence, the nurturant morality, which has a wider sphere of ethical concerns and is more universally applicable. We might think of this in terms of models of moral development, as found in the works of Lawrence Kohlberg or Carol Gilligan, in which the more nurturant model constitutes a higher stage of development. Or we could look at this in terms of political developments, where authoritarian systems tend to precede the development of more democratic systems historically. Over time, the systems tend to become more progressive and caring, insofar as they spend more on say, basic healthcare, family leave, unemployment insurance, etc. Why have you chosen not to use a more developmental model to demonstrate these differences, and why, instead, have you developed a model that could so easily be mistaken as sanctioning the two views as morally equivalent?

The developmental model is simply false, it is just not true. It is a myth put out by people who are authoritarian and in charge. Historians have shown us that both systems have always been there and the reason they have always been there is children have been better off, both when they listen to their parents and when they are nurtured. It is just that these two systems have different social uses and they appear differently in people's brains. But they have been there all along, it is just that the people who have written the histories have been the authoritarians. And when better historians come along, they find out that the developmental story is false.

Now I laid out the model so even handedly because I asked a question that required that answer. I asked the question, "Why is it the conservatives have the collection of views they have?" That is, why are they against abortion and taxation and environmental laws? What does taxation have to do with abortion or environmental laws? And what does being against tort-reform have to do with environmental laws or abortion. And I realized that I had the opposite views on this long list of things, and I could not understand the conservatives, so could I understand mine? And the answer was that I could not - until I realized that one of my students had written a paper on the nation as family metaphor: we have Homeland Security, we have Founding Fathers, who send their sons and daughters to war, and so on.

Then I said, "If this is the case, then we should have different metaphors for the family." And I worked backwards from the politics to the metaphors and those are the two models that I found, and that was the answer to the question. It was a cognitive science study, pure, straight forward, cognitive science. You take this problem, you apply the methodology of cognitive science, you get this answer, you write the book. That was the answer. Since then, deeper questions have arisen that I have had to answer. In the most recent book I published, with Elizabeth Wehling last year, called *The Little Blue Book*, we asked the question, "Are these two moral systems *just* equally?" And the answer was that they are not equally morally valid for a very deep reason: there were deep social truths that were true about one and not true about the other. And that had to do with the fact that in the U.S. and in other European countries, the model that says citizens care about each other and act through their government in making public provisions, like the roads and bridges and public transportation, public science and health and all those things is true. People who started a company and made millions of dollars did not do it all by themselves. They did it with what was provided by their fellow citizens through the government – not that the government did it itself, rather the citizens had to say, "This is what government should do." And that is a truth that contradicts the view that democracy is just about the liberty to do anything you want for your own interest and to ignore other peoples interests, but also that you do not want other people to help you either. That view is simply empirically false about democracy in the U.S. and other advanced countries. And the fact that there is a truth that fits one moral system and is utterly morally false in the other is important.

Tell me about the New Enlightenment and how it might transform our relationship to nature.

If you understand that direct causation is something built into your brain because you are brought up a certain way before you were five-years old but that the world does not work that way, then it is not just the ecological system that is systemic, it is also the global market system that is systemic. Therefore, you have to learn what a systemic cause is and how to recognize it. And then, people have to study and teach systemic causation. And you have to have very special teaching about it as early as possible, because it does not arise naturally. It is something that really must be taught and should be taught as early as children can learn it, because it is absolutely crucial to every part of our existence.

You learn very early in life to think metaphorically and you do not know it, and there are certain metaphors that are learned around the world, because of the way people generally live in the world. You understand achieving a purpose as reaching a destination and that things can stand in your way. You understand quantity in terms of verticality: more is up, less is down. And there are many cases like that, by the hundred, in which there are metaphors that are structuring how you think.

Those metaphors effect policy decisions, and you need to know how your metaphorical system is effecting policy decisions. That does not mean they are all wrong, because metaphors are not arbitrary. There is a reason we learn the systems that we do. But some of them are arbitrary and some of them are disastrous. And you need to know what they are to know which ones are good for you and which ones are not. And it is not just metaphor, it is all framing of issues. You need to know that the frames are there, but they are unconscious. Just simply going along and thinking, "I am conscious, I am smart, I should be able to understand anything," is simply not true. Just because you are conscious, you are smart, and you have been taught how to reason in college does not mean you even know how you think.

Can you talk a bit about Sky Trust?

This was an idea of Peter Barnes. Peter asked the question, "Who owns the air?" and the answer he got was, "We all do." And that is important because oil companies are producing a product that is dumping stuff into our air, and they are not cleaning it up. Normally in a business, if you produce garbage you are expected to clean up after yourself. Then there are businesses that have externalities. They throw their garbage into rivers. They throw their garbage into the air.

They throw their garbage into all kinds of other places. So, if you say, "We own the air," then we have a right to charge for dumping permits. If you are going to dump something into my air, then you should be paying me for that right. Then he says, "If you have dumping permits, what should you do?" You can charge for them. You can say, every oil or coal company has to pay for dumping permits for the amount of carbon dioxide they produce. You can have a minimum charge for a dumping permit initially. Then you can let people trade their dumping permits if they want. But you need to be able to say that you are going to produce fewer and fewer dumping permits every year for, say, 30 or 40 years, where you drop the number by 2 or 3 percent a year for a while.

Meanwhile, a lot of money is being produced. Where does the money go? Not to the government. The money goes to a trust, and what the trust does is it redistributes it to all the adult citizens in the country through transfers to their bank accounts. They do this with three-quarters of the money, and with the rest, they invest in sustainable energy. That is Sky Trust - it is very simple. And a version of this exists in a Senate Bill called the Clear Act. It is there, it is understandable, and if you adopt it, it will clean up the air in 30 to 40 years.

JULIET SCHOR

Juliet Schor is the author and editor of ten books, including her 1992 bestseller, *The Overworked American.* Her 1998 bestseller, *The Overspent American,* was described by the Los Angeles Book Review as a "masterful take on the human folly of overspending" and by Publishers Weekly as "a trenchant commentary on Americans' overspending."

In her most recent book, *True Wealth,* she argues that instead of using productivity growth to get wealthier, we can use it to free up more time. This will reduce our carbon footprints while increasing our sense of well-being. By reducing our hours of work still further, we can create an economy in which more people have the opportunity to work. With more time on our hands, we can make use of some of the wealth we might otherwise possess by participating in various sharing platforms, which make better use of resources. Through all of this, we can make a sizable dent in our greenhouse gas emissions.

Schor is a co-founder of the Center for a New American Dream, the Center for Popular Economics, and South End Press. Before coming to Boston College, where she is Professor of Sociology, she taught in the Department of Economics at Harvard. In 2006 she was awarded the Leontief Prize by the Global Development and Environment Institute.

THEO HORESH: *You have written, "The consumer boom of the 1990s and 2000s was an historical anomaly. Never have so many bought so much for so little." Much of what is driving the unsustainable nature of our economy and lives is still our own consumption. Yet, you seem to imply that something significant has changed since these decades have passed. What do the economic changes we have seen since that time imply for the way we will come to live in the future?*

JULIET SCHOR: The global financial collapse starts in 2007, and hit full force in 2008. It undermines a number of the conditions which lead to the consumer boom prior to it. Number one, it undermines the very cheap and available credit that the financial system gives to consumers. And it does that to keep the system going, because people are not getting wage increases, and yet there is a need for demand, so the financial system pumps out credit. They are making a lot of money on that credit. As long as the system keeps going, we do not have to face the music. So, that is one of the conditions that changed.

The other big part of why that boom happened is the availability of very cheap imports from China, because the boom is in manufactured goods. Those are artificially cheap, because labor is being repressed and environmental costs are not being paid for manufactured goods. There is also very high productivity in manufacturing, but that depends on an exchange rate that is favorable for cheap imports, and it requires purchasing power for all that demand. So, once the economy collapses, consumers are no longer willing or able to take on so much debt.

You have written about this as an opportunity to develop a more sustainable economy. Tell me about that opportunity.

It is an important opportunity to develop more sustainable patterns, because when the crash came, many people felt they were experiencing the hangover phase of a binge. Many of them did talk about it and think about it that way. People felt they were on a kind of consumer binge, and they understood it was not sustainable, because for so many it was driven by credit-card availability. People were taking on debt to consume, and when times got bad, they did not want to do that any more, because it made them vulnerable.

Now you get that natural kind of cautious tendency that consumers have in a recession. People want to pay down debt and build up their savings as much as they can in periods like that. People now want to be more frugal and more cautious, which means lower impact. They are also looking for ways to consume much less expensively, and one of the things about many of the innovations in sustainability, as well as old-fashioned sustainable practices, is that they are less expensive, because expenditure and energy-footprint and eco-footprint tend to go together. For example, you can put your clothes in the dryer or you can hang them on the line. Line drying does not cost you anything in electricity or depreciation of your appliance. And of course, it has a lower footprint.

Many of the things that allow people to save money also lead to lower footprints. Buying used goods is a practice that has gone up a lot since the downturn: swapping, sharing, gifting, and participating in new sharing platforms such as ride-sharing, car-sharing, couch-surfing, etc. Those all tend to be lower footprint ways of getting access to either lodging or transport that are much cheaper for people. So, you have the confluence of doing something in a less expensive way that is also lower footprint. For many people it is also kind of cool, because it is new and it is an innovative way to do something. And that adds a kind of a buzz to it that is appealing for many people.

You have written, "Truth wealth can be obtained by mobilizing and transforming the economies of time, creativity, community, and con-

sumption." This seems to get at the heart of some of what you are saying. Can you talk a little bit more about how transforming these economies, as you call them, can transform the larger economy.

The Plenitude Model, which is the basis of the book *True Wealth*, shifts away from a heavy emphasis on the market, in terms of trying to maximize your earning power and maximize your spending. And it shifts attention to a style of life in which people tend to work less and create and do more for themselves. They may start to grow vegetables, they may join the *maker movement*, they may start a small business, they may get involved in some kind of a community enterprise. There are a range of activities that involve skills and creativity, as opposed to de-skilled work in a bureaucracy. These involve more manual skills and interaction with the material world but also a greater community orientation. This means doing things with other people, or entering into exchanges with people, to build up a community economy or economy of reciprocity, where people are trading things or bartering or informally sharing. I do for you, you do for me. We might even buy a lawn mower together so we do not all have to have them for ourselves. There are a wide variety of local economic interactions that are in the community.

You might be working fewer hours at a job and earning less money, but you have more time. And in that time you do a little bit of self-provisioning. You make and do some things for yourself, so you no longer have to buy. You may make and do some things, with which you earn some money, or you get access to other goods and services. And partly what you are doing through those activities is building community with other people, which is another source of wealth. The idea is that we have wealth in each other, we can depend on each other when times are bad, so it is a less market oriented way of life. It is about a diverse set of activities, and diverse streams of income and access to goods and services.

It seems like a lot of what you are talking about is using the economic recession, and a number of weakening economic trends, to rebuild the social fabric by taking more and more things out of the consumer economy.

Rebuilding the social fabric is important, but I am talking about doing it in a different way than many people have been thinking about the social fabric. For much of the post WWII period, the social fabric was thought of in terms of common interests, common leisure patterns, or common world views, and we make friends on those bases. What I am talking about is returning to economic interconnections at a small-scale, because ultimately it is very hard to sustain community just on the basis of what we might call affinity. And those affinity communities have shown themselves to be weak in certain ways.

By creating economic interdependencies we will forge stronger communities going forward, and we are really going to need that. So, the social connections come along with the economic interdependencies - and I am talking about it at the level of trading. But there is also a lot that is going on since the recession around cooperatives. This includes community land trusts, the idea of investing in our communities, and in groups of people within our communities, in much more egalitarian ways and in ways that really valorize communities, both on the side of producing, as well as on the side of exchanging goods and services and consuming them.

The cooperative movement has been historically successful in a number of places but also very slow to grow. But something has changed in the past 20 years around the rise of the Internet, and especially social media: we now have the capacity to organize ourselves in such a wide array of patterns that we would never have begun to think about before these platforms had arisen. Can you talk a bit about the potential for reorganizing the social fabric, or our basic social structure, around a sustainable economy through making use of some of these tools.

There are various online sharing platforms, many of them in the consumer space, but not only in the consumer space. We are looking at some of the open educational platforms and maker spaces, which tend to be less about the social media. They are really important, because they allow for peer economies to emerge and create much more possibility of egalitarian social relations through those peer economies. They have the potential to be very efficient, along a number of dimensions, and to reorganize the value chain. I have been studying this for the last 3 years through my project on *connected consumption*, which is funded by the MacArthur Foundation. A lot of the things that we spend a lot of resources on in the normative economy, what I call the "business as usual economy," become irrelevant in some of these sharing and peer economies.

The business as usual economy has to devote a large amount of labor and resources to unproductive things, like supervising and excluding people. Enormous efforts are put into those things, and in these peer economies you do not need any of that, so they are wildly efficient in certain ways. They are not profitable in the old sense of the word, in that it is much harder to monopolize the value in these economies in the hands of a small number of people. I would say that is a good thing, though some would disagree. It is a lot harder for the value of profit to be captured by small groups of people in these peer economies.

A good portion of your writings focus on the cultural transformations that need to accompany green-tech innovations if we are to live sustainably. Many of these changes tend to be associated with the hippy movement that has been widely if often unjustly marginalized. And yet much of what you write about, like the potential for small-scale production through 3D printing or the share economy, lies at the cutting edge of high-tech innovation. Can you talk a bit about the interplay between these two cultures of innovation, their values and styles, and what their integration might look like?

It is an on odd juxtaposition in my work in some ways. To me it seems very logical, though. *The hippy movement* of the sixties and seventies got it right on the values and the culture. The return to eighteenth or nineteenth century technologies, which some of these groups were involved in, though, with all of their back-breaking labor, was a big problem, because it was unnecessary. People produced like that back in those times because they did not have other options. We now have the possibility of dramatically increasing both the productivity of labor and the productivity of natural resources. That is the key thing, because we have increased the productivity of labor by trashing the planet and trashing resources. Now, we have to go back and re-do production in ways that get us tech-smart, high-technology ways of doing things that are very resource and productivity enhancing, to use conventional economic terms. We use up fewer resources, we have closed-loop cycles, we are bio-mimicking: all of those things that are in that green-tech space that you mentioned. What this allows us to do is to create those kinds of social relations that these people in the sixties and seventies were looking for, such as peer production, which is very egalitarian. It has sharing at its core, which is, of course, very essential to hippie ideology. It is based on cooperation much more than competition. The new technologies allow us to mobilize those old values, and I think this is a theme that is coming up now in the literature. I did not use this term in my book, because I think there is a lot of negative reaction to the term peasant, but in a way, my model is a kind of *post-industrial-peasant model*, in that you have that sort of diverse set of activities, diverse set of income streams. It is a lot of small-scale, equal producers. But it is post-industrial in the sense that they are using cutting edge technologies.

There is something about your model that is incredibly elegant. You are able to not only create some ideal off in some other place about how to live sustainably in a way that is more efficient, that requires less energy, but you do it from within the discipline of economics. You seize on the opportunity of the recession, seizing as well on the lack of resources, and the need for this transition. Can you talk a bit about the sheer simplicity of what you are arguing for and how many concerns it meets.

It is an attempt to solve two problems at once: one is unemployment and stagnation, and the other is environmental degradation, and climate in particular. The standard approach on climate is technological, and also, for many people, to consume less and grow less. That then exacerbates the unemployment problem. The standard approach on unemployment is to grow as rapidly as possible, and that, of course, exacerbates the climate problem. So, you have two urgent problems. You cannot just go for one or the other, because there is no politics to support one or the other. The constituents for the two need each other to make an alliance, and so environmentalists and labor, broadly speaking, are stuck in a box, where the conventional discourse pits them one against the other.

The core of the answer starts with time use, and that is work that I began back in the 1980s, which is the question of reduced working hours. That is essential to the model. The shorter hours of work are key to reducing those carbon footprints. Since my book came out, I have done some new estimates of the impact of shorter hours of work on carbon footprints and eco-footprints, and it is very large. Countries with short hours of work have much smaller carbon and eco-footprints, and countries like the U.S., which has long hours of work, has other things equal, much higher eco-footprints and carbon footprints.

So, the simplicity comes from that variable, which has big impacts all around. And then of course, we add in that whole analysis of what we do with our time. And that is where we start also to think about this small-scale self-provisioning, what I called in the book instead of the post-industrial-peasant model, *the new economics of household production.*

One of the things we are seeing in our economy is vast disparities in wealth. But corresponding with that we are experiencing vast disparities in time spent at work, with the people at the top not getting the leisure that you would expect to come with greater amounts of wealth. They often put in the longest and hardest working hours. And this seems to be driving a trend towards greater unhappiness and just recently a greater lack of personal health and well-being in general. Can you talk about some of the studies that have been done on happiness, what contributes to greater happiness, and how fewer working hours might affect our levels of happiness?

That maldistribution of hours, some people having too many and some people not having enough, is another important distributional inequity in our system that I am trying to find a way through. It is part of why income is as unequally distributed as it is, because for most people access to hours in the labor market is where their income comes from. We need a way to rebalance that distribution of hours across the population. People that do not have enough hours are unhappy, because they are under-employed. They do not tend to get enough income. The people who have too many hours are unhappy, because long hours of work contributes to unhappiness. There is not a lot of research on this: most of the happiness research looks at things like income, marital-status, health, social connections, and meaning. All of those things have big impacts, but work does as well. The times when people are most unhappy are commuting times, going back to work on Monday morning. You have got all those heart attacks that happen when people go back to work.

So, work time turns out to be an important negative component of well-being or happiness.

How is this likely to affect our climate footprint?

There is a simple way to think about it. We tend to have productivity growth every year in an economy, and what that means is the workers in the economy can produce more with a given amount of time than they could the year before. Now, we can either say, "Great let's produce more," and that will lead to higher carbon emissions. To produce more, we use more energy, more materials, etc. Or we could say, "Let's produce the same amount that we did last year and just work a little bit less. That is what we will use our productivity for." So, over time the decision to produce more instead of working less becomes a central driver of greenhouse gases. The countries that said, "We are going to give people more free time, longer vacations" (in Europe, they are getting five or six week vacations and have more holidays, etc.), are not producing as much as they could. Their people could be working longer and producing more. They are taking more of their productivity dividend in the form of reduced work time or shorter hours, and they have much lower carbon-footprints and eco-footprints as a result.

That time use question ties into growth: how fast does your economy grow? Do you try and grow as fast as possible, or do you use your productivity growth to give people more quality of life and more happiness? The countries that have done the latter, that use more of their economic dividend, more producitivity growth, for shorter hours of work, tend to keep more people employed and to have much lower carbon footprints. That is a very direct connection.

There is one other dimension of it, which is that if you are a household that has a lot of free time, you have the time to do things in lower-carbon-intensive ways. You can take public transportation or longer vacations, and you do not have to jet off for three-day weekends. You can maybe go by car, or you can go by public transportation to places. You can do things more do-it-yourself, which will tend to have a low footprint.

I can think of at least three major ways to reduce consumption. One is voluntarily: we make ethical decisions to consume less meat, to drive less, to live in smaller houses. This is usually very difficult for people to do voluntarily on their own, and I have not seen much success in this area, although it has been an area I have emphasized in my own life and in various campaigns in which I have participated. Another way is from the top down, through say, consumption taxes or a carbon tax. We have these social-engineering-through-tax policies and various incentive structures. But it seems what you are getting at is a third way, which emphasizes our ability to reorganize society and community in voluntary organizations that encourage a shift in our patterns of behavior and thus our patterns of consumption. Can you talk about how that fits in with these other two ways of reducing consumption?

I think I would agree with you that the first does not tend to work very well. You do have a group of people who do that, and some of these things are becoming more culturally popular, so they are spreading. But the key thing, especially for carbon, is how much income/demand is flowing through the system, because if you give people the income, it is very hard for them not to spend it. There are enormous pressures in our society, structural pressures, all kinds of pressures, to ratchet up spending. I wrote about that back in the 1990s. So, the trajectory of shorter working hours pulls that income out of the system in the sense that it is not increasing the incomes that people are getting. That is a really powerful way to do it. What I like about it is that it gives people something that is extremely valuable to them and also creates the preconditions for all these other things we are talking about. You cannot have people doing "making" and small-scale activity, you cannot have them getting really active in their community, and starting land trusts and community consumer coops or any of these things, if they do not have time.

America is now on its third generation of young adults talking about moving beyond industrial development and building a sustainable economy. The results seem mixed at best. What have we learned since the early seventies back-to-the-land movement, what sort of groundwork has been laid for the emergence of sustainable institutions, and what is next?

We have learned a fair amount. The first thing is that the back-to-the-land movement had a certain naivety about what it takes to change a system. The second thing is that sustainability strategies need to be urban in the twenty-first century. So, number one, the rural agenda of back-to-the-land is probably not feasible given the tremendous move of populations into urban areas. And number two, we now know it is much easier to reduce carbon footprints in urban areas, which are dense, have more public transportation, more opportunities for low impact movement, as well as smaller living spaces.

You have written about research indicating that income and human well-being expand when "degraded land, water, and ecosystems are cleaned up and repurposed by people, who live in and around them." How extensive is this research and what does it imply for you?

I was talking there specifically about research that goes under the rubric of natural assets projects done by a colleague of mine, James Boyce, out of the University of Massachusetts, along with a number of collaborators. They have looked at sites all over the world, where poor and low-income people are given access to degraded ecological assets. In the United States, it would be a brown-field, or maybe an inner city neighborhood that has been depleted with lots of vacant lots and polluted areas and a lot of crime or not much going on. In places like India, where you have seen a lot of this, it might be areas of land that are nominally owned by the state. Nobody is using them, because they are degraded in one way or another. These communities are given access to these sites and are able to regenerate them or get involved in ecological restoration of one form or another. And they get one kind of title or another to the land, so they have some kind of ownership claims on the land, which means they can also get the income streams that become possible through this ecological regeneration.

It is a kind of triple-dividend approach. There is the eco-restoration, there is the income generation and poverty allevation, and there is the empowerment. Because in these cases, people come together and democratically decide how to do things and what to do, etc. So, you have democracy, ecology, and economics all moving in the same direction. Now many of us are using the rubric of new economics to describe doing that on a large scale, looking at how to restructure your economic relationships to natural resources, to the earth, in a democratic and collective way, that allows you to yield all three of those benefits.

You have argued, along with other ecological economists, that the discipline of economics needs to incorporate ecological concerns. Which tools and concepts from ecological economics stand the greatest chance of being incorporated into the mainstream discipline of economics, and which do you believe to be the most important?

The one which has been most incorporated already, and stands the greatest chance, is ecosystem services. It is basically the idea of marketizing or monetizing nature. That is totally consistent with the discipline of economics; it is an expansion of the market. Personally, I have issues and concerns with that approach. It is very important to value nature, but when we put a price on things, we also know that the forces for destruction become very high. And this is one of the things that indigenous peoples around the world are so worried about and why they are opposed to the monetization of forests as a part of climate agreements. Their experience has always been that when things get monetized, they get destroyed. So, my own view is that the big lesson mainstream economics needs to learn from ecological economics is not to price everything but that the economy exists within the biosphere, and the biosphere should put limits on the size and nature of human activity. The biggest lesson that mainstream economics should take from ecological economics is the critique of growth.

A world changing idea need not use any resources in its production, and knowledge and creativity are, in general, rather ephemeral. So, it is often argued that as countries grow wealthier, their economies will dematerialize, requiring ever-fewer resources to produce ever-greater value. But you have criticized this idea extensively, and you seem to have some good statistical data to back up your argument. How does this theory of dematerialization play out in the real world?

It plays out in the real world, in part, by the relocation of the material intensive parts of production out of the rich countries. This is also called the *outsourcing of pollution.* North America and Europe look a lot better on their carbon reduction from the point of view of what they are producing than what they are consuming. It is a little bit, "out of sight, out of mind." We have relocated a lot of the really dirty stuff, and energy intensive activity, to the global South. We are doing much less of the primary-sector extraction stuff, like mining and logging, in the rich countries. And of course, we have also relocated a lot of manufacturing. So, part of what lead people to think we were dematerializing was that the extraction and production was relocated as the economy got more and more global.

Can you talk about how some of your goals relate to dematerialization?

The sharing economy could create great opportunities for dematerialization, because it could reduce the demand for new products and new consumption. It can also bring this shift toward shorter working hours that is associated with at least a much slower rate of growth and maybe even with de-growth. So, the material dimensions of the economy could shrink. Digitization itself also creates the possibilities for much less material consumption. Of course, if you take something like paper, certainly in the first stage of digitization, it led to a lot more paper use than less. But I think that eventually some of this digitization is going to lead to those dematerializing effects that people expected. But we have to think about controlling the demand for energy, and I do not think we are going to get to sustainability unless we do something about the demand for energy. We need to talk about that on the way to 9 or 10 billion people, and what the impact might be of bringing all those people out of poverty and so forth. So, the clean-tech shift has to be done within an environment of rich countries really ratcheting down their overall material demands.

Now a lot of environmentalists, and particularly new environmentalists like Thomas Friedman, focus most of their attention on climate change. And it is really easy to roll most environmental concerns into climate change. You have got deforestation and loss of life in the oceans, dirty-polluting industries and fossil fuels, desertification and the hole in the ozone layer. And yet there are so many other environmental issues that can be ignored through an exclusive focus on climate change, and it becomes easy to focus on just a few variables, like CO_2 *and methane emissions, and to ignore all the rest. It seems like you are arguing for a much broader approach. Can you contextualize the importance of climate change in your thinking and how important climate change is relative to other environmental concerns?*

For me, it is at the core of everything. Not absolutely every environmental problem is connected, but all the ones you just mentioned are: deforestation, oceans, etc. Climate change is wreaking and will intensify wreaking havoc on all those other things. I do think climate is the key and that if we are able to solve climate change, the source of a lot of these other pressures is really going to abate. But that means really solving climate change, which means that we have got to dramatically reduce the human footprint on the planet.

Let us say everything is going well with shifting into a share-economy, where we share more of our resources, we work less, we use goods that are less material-intensive, they have lower greenhouse gas emissions, we have solved some of the clean energy problems, and we now have more of a clean energy economy, and not just a lower growth economy, with a stronger community orientation – let us say we have got all of that. What is on the edge of your thinking, concerning the kinds of new innovations in reorganizing society that we might come up with? We have got things like car-shares, and bikes-shares, and we have got Craigslist and EBay and Air Bed and Breakfast. What is the next great innovation in human organization that we need to be on the lookout for?

Then maybe we could get to a post-scarcity economy in the sense that people could be engaged in production as they wanted to, that we could make enough for everybody, and that we could really have an economy of tremendous freedom, rather than coercion and compulsion, which is key to motivating people today. You need a set of economic institutions that make that possible, but basically people could work as much as they want. Their basic needs would be available to them essentially free. And then they could create or do what they like with their time. It sounds pretty utopian. But I think if we could get those footprints down, we could get ourselves into those kind of social relations.

PAUL EHRLICH

Paul Ehrlich is the author or co-author of over 40 books and over a thousand scientific papers and articles. Many of his writing have been in collaboration with his wife Anne and have ranged widely from academic ecology to environmental psychology, from demographics to socio-biology. He is a MacArthur *Genius Grant* recipient, and winner of the Crafoord Prize, a Nobel equivalent awarded once every four years to bio-scientists. He has been a member of the Stanford faculty since 1959, where he is President of the Center for Conservation Biology, and is a past President of the American Institute of Biological Sciences. He has been a major public figure in the environmental movement, sitting on numerous boards and appearing on countless television and radio shows.

Perhaps he is best known for the bestselling classic, *The Population Bomb*, which Stewart Brand has described as one of the greatest self-negating prophecies of history. Ehrlich has argued for almost half a century now that increases in human population contribute to widespread hunger and intensify ecological stresses, and that these problems will only grow as population increases. Prominent amongst these stresses is the rise in greenhouse gases. Stabilizing human numbers will play a major part in solving global warming by diminishing the use of fossil fuels and the destruction of rainforests. But Ehrlich stresses that climate change is only one of numerous environmental threats and should not overshadow the rest. For Ehrlich, the possibility of civilizational collapse is real and his arguments are backed by peer reviewed science. But if we are to take these problems seriously, he suggests, we must extend our empathy to the whole of the world and more closely examine the way we experience environmental threats.

THEO HORESH: *You have received a lot of criticism for what many have labeled a false prophecy concerning the hundreds of millions of people who would starve to death in the seventies. But Stewart Brand has pointed out that this was one of the greatest self-negating prophecies in history. Because of you and others warning of our inability to feed the world, there were significant moves to limit population growth in countries like China, as well as massive investments in agricultural science, with corresponding technological advances, which allowed us to feed billions more people.*

What seldom receives mention was your early prescience on climate change. In a few short pages of your 1968, Population Bomb, you point out that global temperatures were rising due to the carbon dioxide being emitted into the atmosphere but that this had been offset for the past couple of decades by the aerosols and smog that were reflecting much of the sunlight away from the Earth and therefore simultaneously cooling the planet. Hence, we had a general warming trend that was being temporarily offset by a simultaneous cooling trend, both human-induced. This insight is now common knowledge amongst climate scientists, but continues to elude climate skeptics. I am wondering if you could say a bit about the need for environmental forecasting and some of the dangers involved in this tricky art.

PAUL EHRLICH: There are obviously a lot of dangers in forecasting the future at any time. If I was 100 percent good at it, we would not be having this conversation, because I would have bought low, sold high, purchased Bora Bora, and be living there now. First of all, I think it was nice of Stewart to say that, and it is certainly true that a lot of changes came about because of the concern over population and the environment in the seventies. But we should not forget that hundreds of millions of people, several hundred million, have died of starvation and starvation-related disease since then and that we now have close to a billion people who are under-nourished, and some estimates say about 2 billion more who are micronutrient malnourished at a serious level. So, it is hard to be totally cheery in this area when you figure that we are about to add, if we are to believe the projections, about another 2.5 billion more people in the next 45 years – that's half a billion more people than were even alive when I was born.

Pascal famously wagered that in believing in God, he would win whether he was right or wrong, for if there was a God, he would enjoy everlasting life; if there was no God, he would lose nothing. At the end of the Population Bomb, you turned this reasoning on its head, arguing that if you were right about the dire threat of population and we act on this knowledge, we would save civilization; but if you were wrong, we would lose nothing. For human numbers would have to come down sooner or later, and it might as well be now. Have your views on population changed since writing these words in 1968?

Sorry to say, I have become much more pessimistic, because we have not done anywhere near enough. And again, if we are going to have 2.5 billion more people, when we cannot feed the 7.1 billion we have today, that is not a cheery prospect, particularly when you add in the fact that we are not doing anything significant about climate disruption. And climate disruption impacts directly on agriculture; agriculture is responsible for about one-third of climate disruption itself; so it is working against itself. And we are already seeing signs of farmland yields of food, which have been increasing for a long time, now slowing down. And in some places they are starting to drop. So, the overall population picture is one we feared back at the time of the Population Bomb. What is crystal clear is that the population explosion has got to end. It cannot continue forever.

You cannot have infinite growth on a finite planet. You cannot get off the planet. And so the issue is not whether or not the population explosion will end, the issue is how. The two choices are, basically, by a great increase in the death rate or by us doing the things that are necessary to bring birth rates down to the point where we enter a long-term period of gradual shrinkage.

There are numerous leverage points one could target for limiting environmental damage or the potential catastrophic effects of climate change. We could focus on energy issues. We could focus on consumption. We could focus on education. Why have you focused so much attention on population?

The reason is the same one that originally led to the Population Bomb. People were starting to talk about all those other things, starting with Rachel Carson. What was not being mentioned was population size. And what every scientists knows is, the more people you have, the worse all those problems get. And there is no way to solve those problems if you allow the population and consumption to keep growing. So, focusing on population was something we did simply because nobody else was doing it at the time - it was taboo.

It seems this is a really difficult topic for a lot of people to talk about.

It is an indication of how innumerate our society is; we do not teach people mathematics. As the late Al Bartlett kept pointing out, people do not understand exponential growth. And our growth for a long time was close to exponential. Fortunately, it is not now. But the society has got all kinds of forces pushing against recognizing that there is such a thing as too many people. One of those forces is obviously our love of children. Almost everyone likes children, I certainly like them. Some of my favorite times are now with my grandchildren, my great grandchildren, and children of my younger colleagues. The idea that there can be too many of them is difficult for many to understand.

Then there is a lot of pressure from some religious and other groups that want to outbreed still other groups, thinking that the number of people in their group is a source of strength. And there are also all kinds of related equity issues. For example, nowhere in the world are women given equal rights to men. As you increase the rights and opportunities of women, birth rates go down, and that is the ideal place to start bringing birthrates further down. They have come down in much of the world, but we need to bring them down further so we get a gradual shrinkage of population back to a number that can be sustained in the relatively long-term.

It is now common wisdom that the best way to limit population growth is to educate girls, but good K-12 schooling seldom appears until a country becomes somewhat developed. Building decent schools costs money, as does educating teachers. Often times, just getting money to schools, as opposed to corrupt elites, requires democracy. But it is often difficult to build a democracy when a country is very poor, because officials tend to be more corrupt. Then we have this problem that the most developed states tend to fair far better than the least developed in limiting their numbers, and yet they have these tremendous ecological footprints. How might we escape these paradoxes.

We now know that you do not have to develop a country in order to produce education. If you only introduce contraceptive technology and knowledge to a society, it tends to increase the education level. And you do not have to have a super rich country, with high levels of consumption, in order to bring birth rates down. In India, for example, in the state of Kerala, the total fertility rate is now about the same as the United States. Through basically the action of one person, introducing condoms on a big scale. Thailand brought their birth rates way, way down. It has happened in a lot of poor countries, and it really does not involve full-scale industrial development, as occurred in the rich countries.

But as you just implied, the most important place to get birth rates further down is in places like the United States. We are the most over-populated nation in the world. We have the third largest number of people. But when you multiply in the consumption patterns in the United States, we are still the leading over-populated nation. Nobody has ever come up with a sensible reason why there should be more than 120 million Americans, and the only reason for having 120 million is some people think falsely that numbers are the key for military power. We won the greatest war in history with about 120 million people. Now we have over 300 million, and that is many more Americans than we should have alive at one time. If we want to maximize the number of Americans, we should do it by having a sustainable number over hundreds or thousands or millions of years, not seeing how many you can jam into the United States in this century, so the society collapses. Of course, the big issue in the knowledgeable scientific community, is whether there is any real chance of avoiding a collapse of civilization on the route we are now on.

Many environmentalists are uncomfortable with trying to lift the billion or so malnourished people living in the world today out of poverty, because they believe that if we save their lives, they will have even more children, who will themselves go hungry in the next generation. Yet, it is the poorest countries that are populating the fastest, while in many of the most developed countries population growth has leveled off, with the United States being a slight exception. Wishing the poorest people in the world dead can also border on the genocidal and this pits proponents of such views against the neediest people in the world today. You are often grouped in with this set of thinkers, but I read something very different in your writings. You seem concerned about lifting the bottom billion out of their poverty. It even seems like this is a major motivating factor in your work. How might we limit population growth in the poorest parts of the world while extending to the people living in those places our full empathy and care?

I do not like anybody who is already living on the planet having to live a miserable life, exposed for example, as the poor people of the vastly over populated Philippines were, to the recent typhoon. We need to take care of everybody we have got on the planet, with a decent diet, a decent education, and so on. All of which could be supplied even to today's numbers, at least temporarily, if the rich countries would stop consuming so much and share with the poor countries.

If you just dropped what we spend on military adventures, you could bring much, much better lives to the bottom 3 billion in the world. Remember, it is about 3 billion people we have living in misery. And as far as their population situation goes, it is mostly sub-Saharan Africa, where the very high birth rates persist in really poor countries. The world needs to do a lot of work there to give those people the health care, the food security and so on, that they need in order to be able to cooperate with the rest of the world decently in solving our overall problem.

I hate to say to it, but we do not have a prayer of solving this problem, in my view, without some redistribution of wealth. We should be able to have the kind of redistribution in which the rich realize that their fates are tied in tightly with the fates of the poor and begin to give up some of their over-consumption in order to help the poor, but that is a big program. The political and economic barriers to it are horrendous. And of course, the whole economic system is based on continual growth. Even our money supply, with the fractional reserve system of creating money, depends on keeping everything growing. And you cannot keep everything growing forever. But of course, many economists, and most people, just do not get it.

You have pointed out the need to become one human family, to identify ourselves with the whole of the human race, if we are to solve problems that are confronting the whole of the human race.

We are small group animals still, culturally and possibly genetically. We were hunter-gatherers for millions of years of evolutionary history. We have only been agriculturalists for 8 or 10 thousand years, and not everybody for that length of time. And it is thought that the group sizes of hunter gatherers were somewhere between 50 and 150 people. That was the in-group, that was the people you related to, that was the people whose faces you recognized. They all spoke your language, they looked a lot like you. They were genetically related to you. We are used to living with groups of that size. And in fact, when you look at things like Christmas lists and who comes to family reunions, you still have groups of about that size. But unhappily, we have now got 7.2 billion people on the planet, and we have got to learn to deal with bigger sized groups. And we have started in that direction.

If we had a thousand years I would be much more optimistic, because after all, we got rid of slavery over much of the world. We are slowly beginning to give women decent rights. One of the few advantages of getting older is I can remember clearly my first girlfriend in the 1930s, when I was 7 or 8. The only real opportunities she had in life were to be a nurse, a secretary, or a teacher in the lower grades of school. And people with dark skins could not play sports, except for boxing, and they were not in movies, except as janitors. Yet, all that has changed, in no small part by the way of the Second World War. It has not changed enough, but you can see that in less than 100 years, we have made considerable progress.

The trouble is we do not have another 100 years to change the way we relate to people, to change the way we think, not only of the people that are distant from us in space, but also our descendants that are distant from us in time. We have a very hard time relating to any of our descendants that we do not know personally. We cannot picture their faces, and so right now we are using up resources and putting the world on a track that is going to make it very difficult for most people's grandchildren, and certainly beyond that to great-grandchildren, great-great-grandchildren, and so on. We are stealing from future generations right now. It is not even clear the world will hold together well enough for there to be substantial future generations.

The last couple of decades have seen a groundswell of studies on the human capacity for empathy. How is this relevant to our response to climate change?

Empathy is being able to put your feet into other people's shoes, to not just sympathize with them, but actually put yourself in their place. Again, going back to future generations, my interest in environmental problems was very, very high when our daughter was 6 or 7 years old. We brought this helpless thing into the world, and the world was going in the wrong direction. Think about what it is like for a kid today, growing up with the prospect of not being able to eat the way we eat today, not being able to live in many places where we live today. Remember, we have coated the world with air conditioned structures in places where people really cannot live without air conditioning; and yet it is very likely that we are not going to be able to use the energy to air condition large areas of the world any more. If my grandchildren live in say, southern Florida, I know they will, in their lifetime, probably have to move because of the problems of heat and the fact that southern Florida is going to be under water soon. Learning how to care for what is happening to someone who is in, say, Nigeria, requires a great increase in empathy for most of us.

You have been at this a really long time. How do we do it, how do we extend this sense of empathy to include the whole of the human family?

Sid Caesar was one of the early television comedians, and he often played Professor Von Stoopnagle. And they would interview Professor Von Stoopnagle on various things. One day they interviewed him on mountain climbing, and how he climbed sheer cliffs. The interviewer said, "Professor Stoopnagle, what happens if you slip, and you fall off the face of the cliff?" And Caesar sticks his arms out and starts moving them up and down, and says, "Well you know, that is what I will do." The interviewer says, "Professor Stoopnagle, man cannot fly." And Caesar says, "Who knows, I might be the first." Well, that is what I feel I am doing. I am sticking my arms out and moving them up and down.

There are a lot of people concerned with this sort of thing, but what is often called civil society is not unified and organized. For example, every outfit that is concerned with the rights of women, with the rights of minorities, with not having the U.S. be a vast, war-like aggressor state, and so on, should be concerned with population. They should be concerned with the environmental situation. If we have, for instance, another nuclear war, which could easily happen between India and Pakistan, the most recent research shows it would end us too. So, we ought to all be concerned about the really basic issues, and working together on them, and we have not organized that yet.

You have pointed out that our senses evolved to alert us to changes in the immediate environment, the sound of branches breaking or the scattering of surrounding animals. But we can neither see nor hear the emissions of methane and CO2, and since once released these gases spread through the atmosphere, we cannot perceive the damage they do either. Nor can our senses do a very good job of perceiving changes in climate, since weather is constantly shifting, but changes in climate occur over decades, if not centuries or longer. What does this imply for our understanding of climate change?

Our nervous systems were designed to hold the environmental stage on which things occur as constant as possible so that you could detect signs of danger. The job of *Australopithecus* was to duck the leopard, not to deal with climate change, because it had not caused climate change, and there was nothing it could do about it either.

If somebody throws a rock at your head, you solve a whole series of differential equations instantly and duck. That is easy, you see it against the background, you know exactly what to do. The problem is the threat today is not leopards jumping out of the bush generally, or somebody throwing a spear at your head. The threats today are changes in the background, changes that we are designed not to perceive - quite the opposite. Our whole nervous system goes in the wrong direction. The things we have to worry about are the slow accumulation of greenhouse gases in the atmosphere, accumulation of toxic chemicals from pole-to-pole, accumulation of nuclear weapons, accumulation of at least the possibility of being able to design biological weapons that will be truly horrendous. Those are the things that we should be paying attention to, but unhappily, you need real education and training to do that. A chart of the increase of CO2 in the atmosphere, measured at a research station in Hawaii, does not mean anything to the average person. Years of teaching have shown me that it takes a long time to train people to understand charts and graphs.

The big problem we are facing is how you get people to understand the basic facts of how the world works. Almost nobody in our Congress understands, for example, the climate situation. I think most of them believe that if we decided to move to wind turbines for our energy tomorrow, that in two years we would be getting half our energy from wind turbines. They have no idea of the scale of the problems. They have no idea of the time it takes for us to do things, to build new energy infrastructures and so on. We have a huge educational problem, and I have spent a lot of my time trying to figure out how to communicate with people the scale of the problem. We should have been working on our energy system and our population situation 50 years ago, really hard. And every day we waste, we lead ourselves towards a world that could easily become uninhabitable -

at least in most areas.

How do we train people to develop the slow reflexes needed to understand the deep background changes that are threatening the survival of human civilization.?

We have proven beyond a shadow of a doubt that telling people what the science says does not change their behavior. The climate disruption situation shows that dramatically. A few highly paid, intellectual prostitutes are able to counter the entire climate science community's view of what the future holds. And so I am thinking more and more, as are some of my colleagues, that we need a quasi-religious change, that actually we need revitalization - just change the whole system. Sooner or later, we are going to have to change. The issue is whether there is any chance of changing soon enough.

There is a school of environmental thought that suggests we need to get more in touch with our senses so as to better connect with and care for nature; but as you have just been pointing out, our senses are an unreliable mechanism for detecting environmental challenges like over-population and climate change. How can we relate with nature in such a way as to deepen our sensory connection and care, and also better grasp these sorts of long-term changes that tend to require careful and often statistical thinking?

Again, we are moving in exactly the wrong direction. We are becoming more urbanized. There was a time when most people not only understood better where their food came from, for example, but were actually familiar with nature instead of familiar with video games and video presentations of nature. One of my colleagues here is doing research, which suggests that our cognitive functioning improves if we have more exposure to nature as opposed to the environments we tend to immerse ourselves in. But these are the big questions. Academically, the action has shifted to the social sciences. We know more than enough about hard science to understand the trouble we are in and to know the direction we ought to be moving. What we do not understand enough is how you change human behavior. In fact, we do not even take advantage of the things we do know about human behavior.

You have pointed out that we have all these different levels of human organization. In terms of cultural evolution, we go from band to clan to tribe to nation. And we have human families, organizations, institutions, and nations. And you have pointed out that we need to add one more level of organization onto all of this: the global human family or global civilization. But you have pointed out that all of these other levels of organization are important as well. I am wondering if you can say a bit about the integrity of all the different levels of human organization that we need in order to make a global civilization work.

We are failing at many levels of governance, but I think the most serious failure today is the global level of governance. In other words, people seem to think that the nation state is the permanent solution to that problem. But of course, in their present form, they are only a couple hundred years old, and they are miserably failing.

You have noted that John Stuart Mill is often said to be the last person to have known all there was to know. Whether or not this is true we can set aside; but now even very educated people do not understand vast swathes of the world around them. We do not understand how the things we encounter in our everyday lives work. We do not understand the implications of our actions. There is a sense that even highly educated people do not understand what they are doing to the world. And it seems to me that one of the greatest impediments to thinking globally is the sheer immensity of the task. Throughout this dialogue we have bumped up against the limits of our capacities to conceptualize a world, to conceptualize how we will organize at a global level, to conceptualize how we solve problems at this global level. How do we go about thinking of something so vast as a world?

Anne and I have written a lot about what we call *the culture gap*. When I was a kid, I lived with the Inuit, with the Eskimos. And every Inuit basically knew the entire Inuit culture, the entire store of non-genetic information of his or her tribe. The men knew how a woman's knife, an *ooloo*, was made and used. They did not normally use them, but they knew how to, and the women knew how to hunt seals on the ice, even though the men did it primarily. Now we have a situation where, when I talk to an audience of 500 Ph.Ds, which I do occasionally, the whole audience does not have 1-millionth or 1-billionth of the non-genetic information humans as a whole possess.

You can see a sign of the problem looking back at the debates between the Federalists and the Anti-Federalists. The people debating then all had the same base of knowledge. They all knew about Montesquieu. They all knew the classics and what had happened in Rome. And now we have the problem you just outlined. My solution to it would be to totally reorganize the schooling, so there was actually a curriculum, particularly in the universities, that gave people the key parts of their culture. Everybody ought to understand the second law of thermodynamics, because it operates in their lives all the time, and it is critical. Everybody should understand how the climate functions. Everybody should understand demography. Everybody should understand the resource situation and how the food system operates. It does not take that much to teach people the parts, the bridges over the culture gap, that are necessary for everyone. I suspect that one quarter of a 4-year college degree devoted to just that would be enough.

We have at Stanford the best research on interdisciplinary environmental issues, I think, in the world. It is because Stanford has hired a lot of smart people, and smart people understand disciplines are a crock now. They were developed by Aristotle and formalized by the Royal Society in about 1590. So, the really smart people do not pay any attention to the structure but work with each other on these problems. But all of the perks, all of the power and so on, still runs down disciplinary lines at Stanford and other universities. In other words, we have become the best environmental research organization in the world, despite the university's structure. As someone put it, people have problems, universities have departments.

Over the course of our interview we have recapitulated a lot of the issues that have arisen in your career, starting with population and moving more and more towards this social science view of the human animal and our capacity to deal with some of the problems that you first worked on. And this raises a question about how we educate people. You have written more alarmist books like the Population Bomb. You have also written very academic treatises on ecology. And you have written more social science type works on the human capacity for empathy and global cooperation. And some of the people that are in this book, have been quite critical of a more apocalyptic kind of environmentalism, the doom-and-gloom warning of the dangers we are confronting, because people can become so burnt-out hearing this message. The message quickly wears thin; it also invites a kind of manic reaction against it. Tell me what you have learned, over the course of your long career, about the kinds of messages that are going to wake people up and sustain their interest and activity in addressing concerns like climate change.

The main thing I have learned, empirically, from having run the tests writing books warning about what is going to happen and writing books on how we can deal with it, is that if you tell people it is in the fan and you are in deep trouble and you better do something, it has a much bigger impact than saying, "We have problems, but we can solve them, and here is the way we can solve them." All of our solution-oriented books have gotten little attention. The sales of the Population Bomb were probably 50 times that of any book Anne and I have written on solutions – on how to deal with the issues or the technical side of our dilemma. That is sad, but it is, unfortunately, the way the world is. I do not talk stuff that my colleagues do not agree with. My biggest critic on the issue of collapse of civilization is the world's expert on bio-geography, Jim Brown, an academy member with a deep interest in energy-return-on-energy-invested. And I have said to him that I think maybe we have a 10 percent chance of avoiding a collapse, but I am willing to work hard to make it 11 percent, because I have got grandchildren and great grandchildren. And he says, "Paul, you are out of your mind. There is only maybe a 1 percent chance, but I, Jim Brown, am willing to work to make it a 1.1 percent chance for the same reason." In other words, I am one of the least pessimistic of my knowledgeable colleagues.

SECTION III

JOHN BROOME
KATE PICKETT
ROBERT HENSON
HOLMES ROLSTON III

JOHN BROOME

John Broome is the author of seven books on ethics and economics. He is a lead author of the 2014 Intergovernmental Panel on Climate Change report. He received his doctorate in economics from the Massachusetts Institute of Technology. And he is professor of Moral Philosophy at Oxford University. Broome excels in bringing to the field of ethics the insights gleaned from a career teaching and writing on welfare economics.

Lord Nicholas Stern, former Chief Economist of the World Bank and British Chancellor of the Exchequer, has described him as "one of the most outstanding moral philosophers of our time." And the renowned ethicist Peter Singer has described his *Climate Ethics* as "practical ethics at its most brilliant and most significant."

Broome argues that climate change is a moral problem, but one that only governments are in a position to solve. Governments should aim to make the world a better place by limiting the harms of climate change, and the policies that achieve this will need to be founded on ethical principles. Individuals also have moral duties towards climate change, which involve not harming others through the emission of greenhouse gases, something that can be achieved by offsetting emissions. Yet, since it is governments that are most capable of solving climate change, we have a further duty to try and move our governments to act as they should.

THEO HORESH: *In what sorts of ways might climate change inflict harm, and in what ways are we as individuals responsible for the harm that is inflicted through our emission of greenhouse gases?*

JOHN BROOME: You ask how it *might* inflict harm, but actually climate change is already inflicting harm on a lot of people. Most clearly, it is already doing serious damage to the culture and the way of life of Arctic peoples, because the Arctic is where climate change has up to now been most dramatic. But also all over the world it is increasing the frequency of storms and extreme weather of other sorts. It is leading to more droughts, to floods, violent winds, and heat waves. It is causing sea levels to rise, because the oceans are also getting warmer, and as they warm their water expands. The rise in sea level will drown coastal areas – in due course, big areas where millions of people live, such as the river delta in Bangladesh. It will damage water supplies, especially by melting the mountain glaciers that feed water to a large fraction of the world's population. It is going to make farming more difficult. Partly it will do so by drowning farmland, but rising temperatures will also by themselves make farming harder in many parts of the world. Very large areas of farmland will turn to desert. Climate change will increase the range of some diseases – particularly tropical diseases – that will kill many people, including particularly children. It will push many people further into poverty. Poverty leads to malnutrition and consequently to many deaths: within a few decades we can expect hundreds of thousands of deaths per year. It will do very severe damage to the natural world; it will cause the extinction of many species. Acidification of the oceans is also doing a great deal of harm in the oceans, and so on. Climate change will do serious damage to the geography of the world, and have consequent effects of all sorts everywhere.

That is quite a list.

I could have gone on longer, and I need to stress that it is human beings that have done this. The increase in global temperature is the result of human activity.

The author, Steven Gardiner, has characterized global warming as a perfect moral storm, wherein we have several ethical problems coming together that we really have not thought through much before. We have not done a lot of thinking about our responsibilities to future generations. We have not done a lot of thinking about our responsibilities to all people across the planet, and we have not done a lot of thinking, until quite recently, on our responsibilities to non-human life. And so with climate change, we have these issues all coming together, and it happens in this peculiar way, where it is through day-to-day actions - things like driving, eating a meal, and heating our homes - that we are causing all of this harm. Can you speak to some of the challenges involved in confronting so many problems, as a philosopher, upon which there has been so little previous thinking?

It is obvious that climate change is a moral problem. The practical question raised by climate change is what we ought to do about it, and what we ought to do about it is principally determined by our moral duties towards other people. We ourselves are not going to benefit very much from whatever action we take. We will do it for others: for future generations, and for people living elsewhere in the world. So, it is certainly a moral problem. But just because it is a problem on such a very large scale – it affects everybody in the world, and will affect them for centuries – it is going to require the methods of economics to work out in detail the right response to it. A lot of economics is concerned with morality, with what ought to be done. That means that it ultimately has to be founded on ethical principles, but the working out of those ethical principles often calls for the complex quantitative methods familiar in economics.

Some of the questions raised by climate change are actually relatively familiar within the ethical branch of economics. Some of them are about the distribution of wealth or the distribution of well-being, both between generations and between people within a generation. Economists have thought for a long time about how to evaluate distributions across the population of the world and across time. So, to some extent the ethical problems that climate change raises are versions of questions that are relatively familiar to economists. Some of them are fairly new to moral philosophy because moral philosophers have generally thought about small-scale moral problems involving personal relations among small groups of people, but they are not new to the ethical branch of economics known as "welfare economics." I do not want to say that the economists have the right answer to them. Sometimes they do not, but so long as they recognize that the work that they are doing is founded on ethical principles, they can make very useful contributions to approaching the moral problems raised by climate.

That is how I first came to the subject when I was an economist. But it is also true that there are very serious moral questions raised by climate change that economics has not confronted in the past and is not very good at. You mentioned one: how to think about the value of nature. Another is how to think about the value of the Earth's human population. Undoubtedly, climate change is going to affect the population. If we have serious climate change, it may even cause our population to collapse to a much smaller level. It is an ethical matter how we should judge the goodness or badness of a change in populations. Few economists have given much thought to that question, but in the last few decades, the ethics of population has been a significant issue in applied ethics. It is a very difficult one, and in several decades we have not converged on a consensus about answers to it. So, that is a serious challenge that climate change raises for moral philosophy: how to evaluate changes in the Earth's population. And given that at the moment we do not know how to do this, population ethics raises a further problem for moral philosophy: how should we act when we do not have any confidence in any particular moral theory that tells us how we ought to act?

Another important challenges within moral philosophy is how to accommodate the well-being of animals and what to do about the value of nature.

This raises an interesting question. We often hear about how the models projecting future changes in climate just cannot capture enough information. The climatic system, the atmosphere, the oceans, the whole carbon-cycle, is just too complex to be captured by the computer programs that we have today. Climatology has in many ways driven the need for more computing power. But it also seems that something like this is happening with our economic models. There are just so many changes that we need to capture in these models, including the value of non-human life, including aesthetic values. Are you hopeful about the ability of these economic models to capture so much, not just quantitative but also qualitative information?

No, and I do not think economics should try to do that. Economics is useful in one particular department of morality: it is useful in evaluating human well-being, and not elsewhere. It has little to say about the value of nature, or aesthetic values. When it tries to deal with these values, it does so badly. It would be a bad mistake to think that we could extend the methods of economics to cover everything that we need to think about. We have to recognize that economics is seriously limited in what it can do. It should stick to human wellbeing.

But in what is called *aggregation* it does have special methods that are useful. Economists are good at thinking about how we should evaluate large groups of people together, and large swathes of time together, and how well-being is distributed across people and spread over time. That is the special business of economics. It is very important within climate change, because well-being encompasses a lot of the values that climate change will affect. I told you about all the harm to human well-being that will result from climate change. It includes, for example, the shortening of many people's lives. Climate change will kill many people. What is wrong with killing a person is that it reduces the person's well-being. So, the badness of the killing is something that could be incorporated within economics, but there are many things that cannot be.

You have written about the difference between justice and goodness. Why does this distinction matter to the way we think about climate change?

Intuitively, it is quite easy to see that we do make a difference between duties of goodness and duties of justice just by thinking about one classic example that appears in lots of literature of moral philosophy. It is the example of a transplant surgeon, who has in the hospital five patients, each needing a different organ. One needs a heart, one needs a liver, one needs a kidney and so on. They will all die if they do not get organs. The surgeon takes an innocent visitor to the hospital, kills her, extracts her organs, and distributes them to her five patients, thereby saving five lives at the cost of one. So, we now have five people living rather than one person living. That is beneficial in the aggregate. It increases the overall goodness of the world, because five people survived rather than one. But it is obvious that the surgeon should not do this. Why not? Because we must recognize a principle of morality that says you may not harm people even for the sake of promoting the greater good. The surgeon violates this principle. For the sake of promoting the greater good, she kills one person. That is not morally permissible.

So, there is a principle of morality that can even override the moral demand to promote goodness in the world. I treat this as a principle of justice. It is a commonsense principle that we commonly recognize. Roughly it says, "Do not do harm, even for the sake of promoting the greater good." There are certainly limits on it; there are cases in which it is permissible to do harm. For instance, you may do harm in self-defense, but there is definitely a general principle that says you may not harm other people except in special cases.

This principle is important in climate change, because when we cause the emission of greenhouse gas, we harm other people. I described all the harm that greenhouse gas does. Each of us contributes to it. And the amount of harm that we each of us do is roughly proportional to the amount of greenhouse gas we emit. There is no justification for doing this harm; we should not do it. That follows from the commonsense principle of justice I described. It seems an inescapable conclusion that we should not harm other people through our emissions of greenhouse gases. We should stop doing so. That conclusion is an important consequence of the principle of justice within the morality of climate change.

But I want to emphasize that I do not think this is the most important thing to say about climate change. Recognizing that we should not cause emissions, because it is against this principle of justice, is important in individual morality, but the problem of climate change is not going to be solved by our private individual morality. It can be solved only by action on a large scale, by governments and the international community. Governments and the international community should aim to promote goodness, more than justice. They should aim to make the world a better place. Climate change is making the world worse than it otherwise should be. It is the role of our governments and of our international system to prevent this happening, and the purpose of preventing it is to improve the world. It is to promote the aim of goodness, rather than an aim of justice. So, I think justice is important in individual private morality, but what will solve the problem of climate change is the morality of governments, which should be aimed principally at making the world better.

You said our emissions cause harm, and they are going to be causing harm not only to people all around the world, but they will be causing harm to all people in future generations in all of the ways you mentioned in that incredibly long list of hazards. But the idea of getting our emissions down to zero will strike many people as absurd or simply impossible.

I agree: when I say that, at first it sounds ridiculous. But actually, we can as individuals reduce our emissions to zero by means of *offsetting*. Offsetting means that, when you emit some greenhouse gas, you make sure that you also subtract the same amount of greenhouse gas from the atmosphere, so that the net change you cause is nothing. You do not add any greenhouse gas to the atmosphere. This is something you can achieve. You can achieve it directly if you have a bit of land, by planting trees. As trees grow they take carbon out of the atmosphere, and use it to build their bodies. Plants take carbon dioxide from the air, break it down into carbon and oxygen, release the oxygen and keep the carbon to grow their bodies. So, in principal you can offset your emissions directly, but planting trees is probably not a very effective way of doing it. In the end the trees will die, and their carbon will be released back into the atmosphere.

A better way is what I call *preventive offsetting*, which is to prevent an amount of carbon dioxide that would have gone into the atmosphere from actually going there. As you cause the emission of some carbon dioxide, at the same time you prevent carbon dioxide from some other source from being emitted. That is an effective means of offsetting. You can do it by employing an offsetting company to do it on your behalf. There are many offsetting companies that run projects designed to prevent carbon dioxide from getting into the atmosphere. For example, they build sources of renewable energy: hydro-electric stations, windmills and solar-power stations, for instance. Another example is that they provide people in Asia and Africa with efficient wood-burning stoves. Surprisingly perhaps, a lot of the emissions of carbon dioxide are caused by people burning wood for cooking. If only they had more efficient stoves, they would burn much less wood. And several offsetting companies supply efficient stoves as a means of reducing emissions.

It seems that environmental harms are particularly difficult to calculate. We know we are committing great acts of harm through the resources we use, and this can lead some people to a paralyzing sense of guilt. Your effort to account for the injustices inflicted through our CO_2 emissions provides a simple solution and hence moral clarity, but it seems we also need to account for our use of ground water and minerals, the harm we inflict on animals, and the ecosystems we destroy. But this would present a much greater burden of responsibility. How might we begin to think through these resource-injustices that have become so much a part of our everyday lives?

You should not assume that they are injustices. They do harm, or they may do harm in the future, but you said some of the harm is to ecosystems, for example. I do not believe you owe any duties of justice to ecosystems. I do think it is a bad thing if you damage ecosystems, but I do not think that this is a failure of justice. A characteristic of justice is that it implies right, but it is implausible that ecosystems hold rights against us, which we infringe when we harm them. Duties of justice are normally owed to other people. There has to be a right-holder for justice to be at issue, and I find it very hard to believe that there is a right-holder in the case of damage to ecosystems.

I am stressing this because I think it is a mistake to think all of the problems of climate change or the environment in terms of justice. Many of them are matters of the world's being less good than it otherwise would have been. People sometimes suggest there is injustice when there is not, because that is a rhetorically effective claim. If you can suggest that some right is at issue, or that an injustice has been done, it pulls hard at people's moral intuitions. But we are under a different, very strong moral duty to improve the world and not to damage it, and that is the reason why we ought not to damage ecosystems. It is not a matter of justice.

That is the beginning of how we should think about these things: do not assume that everything is a matter of justice. Think instead about our duty to promote goodness in the world. But let me say, the questions you have raised are big and difficult. We damage the world environmentally in many different ways – not just through climate change. And I have not arrived at conclusions about the all moral issues in which we are involved with the environment.

Early adopters of sustainable behaviors tend to suffer far greater challenges and risks than late adopters. The first vegetarians risk social ostracism, while those of today can find aisles of veggie-snacks. The first bicyclists in New York City risk life and limb. Now they can enjoy rides along tree-lined paths. Path-breaking environmentalists make it easier for the rest of us to adopt sustainable behaviors and demonstrate that legislation is often far more feasible than at first supposed. This makes moral commitments seem more meaningful than some measure of CO_2 being kept out of the atmosphere would suggest. What sorts of issues arise in trying to account for these efforts and the benefits they yield?

As I said earlier, I do not think that individuals will solve the problem by fulfilling their duty of justice to stop their own emissions. It would solve the problem if everybody did it, but I am not naïve enough to believe that everybody will do it. So, I think the problem is going to have to be solved through political action. It is going to take governments using their powers of coercion to prevent people from emitting greenhouse gas. Governments can use regulation to reduce emissions, or they can alternatively make it costly to emit, through taxation and other means. So, political action can be more effective in actually solving the problem of climate change than individuals' conforming to their duty of justice by not causing emissions themselves. The path-breakers you mention do just that. They have a political effect. Their own individual actions do indeed fulfill their duty of justice, but because they are showing the way to many others, they also have a much wider effect through the social system and ultimately through the political system. That is why I think their work is particularly creditable. It is opening the way for the political system to start doing something about climate change.

Now, many American environmentalists have come to feel political action to slow global warming is hopeless, so they invest their energy instead into effecting change through private ethical actions and small-scale community projects. How should we think about this sort of response to global warming?

It is very commendable. As I say, I do not think that people should be emitting greenhouse gas, and they are avoiding doing it. But on the other hand, it is not by itself going to lead to a solution to the problem. It really does take governments – maybe not necessarily national governments, but institutions with the power to compel people to reduce their emissions even if they do not want to. Otherwise enormous numbers of people will carry on emitting profligately even when others reduce their emissions. If we all became ecologically virtuous, that would solve the problem, but I do not believe we shall. Therefore, we have to find ways of getting our governments to do something about it.

I started noticing something around perhaps 2003: A wide array of environmental concerns that I held, and that I had often worked on, were beginning to be subsumed by climate change. Not only that, they were beginning to be overlooked, but they could all fit under a concern with global warming. I wonder how much of our environmental concerns we should subsume under the category of climate change. It seems to include a large part of the species loss, a large part of the depletion of ocean-fisheries, coral reefs, and life in the oceans, the destruction of rainforests, etc. Greenhouse gas emissions are tied up in air-pollution, water-pollution, acid rain, and ozone depletion. Should we just focus on climate change and greenhouse gas emissions? How much more should we be looking at all of these other environmental issues?

A lot of environmental issues tend to conflict with tackling climate change. For example, a lot of environmentalists are anxious about nuclear energy, for the good environmental reason that it will leave dangerous nuclear waste lying around on the Earth for many tens of thousands of years. Lots of environmentalists are not happy with the building of wind farms and solar farms, which are destroying the natural beauty of some parts of the world. Climate change has taken a knock recently in Germany when the Germans decided to phase out their nuclear program, as a result of the Fukushima disaster. So, there is a fault-line within environmentalism. The answer to your question is therefore that the concern with climate change does not encompass all of environmentalism. Other environmental concerns are still important.

Now, it is possible that global warming will be catastrophic to all life as we know it on Earth. What is the likelihood we will experience some kind of catastrophic global warming sometime in the next century, and how might we go about thinking about the risk?

The degree of likelihood is not my subject. The scientists do their best to estimate how much the Earth is going to warm, and to figure out what the effect will be on human life. They are beginning to attach some degree of likelihood to extreme climate change. For a long time, they concentrated on what they thought most likely to happen. Now they are beginning also to give us an estimate of the probability of, say, 8 degrees of global warming. But they have not reached the point where they can really say anything very significant about the probability of catastrophe; it is outside the range of their data. We can learn from them that there is a risk of catastrophe, but not much more than that. We can perhaps also learn that it is not a minute risk. It is not, say, one in a million or one in ten thousand. The risk of catastrophe cannot be ignored. That is the best I can say about its likelihood.

The job for philosophy in thinking about catastrophe is to investigate how bad a catastrophe would be. What sort of catastrophe are we thinking of? One that would certainly do a lot of damage to human and animal life. It would leave the Earth with a lot fewer creatures living on it. It might even cause the extinction of humanity. It will certainly cause the extinction of a lot of other species. We need to think about how bad those possibilities would be. What is the value of having people and other creatures living on the earth? That is not an easy question to answer. It is part of the population problem that I was describing earlier. What value do we attach to the number of people who live on the Earth? Would it be a bad thing if there were no more after say, 400 or 500 years from now, and if so, how bad? It is a job for philosophy to evaluate the catastrophe. But to say what are the chances of it is not my job.

Now, this problem of how to think about an ideal human population is enough to make one's head spin. The concern of environmentalists, who would like to ultimately limit human population to say, a billion or even half-a-billion people in their ideal scenario, seems to be with the sustainability of human populations. But it is also with the well-be-

ing of thriving ecosystems, which means in terms of well-being, more thriving plant and animal life. So, when I start to think about ideal human numbers on the planet, I am not just thinking how much human well-being we might get, because that might mean much, much higher numbers of people, than I am thinking how much that could be sustained over time. And I am also thinking what kind of human numbers are in accord with a thriving, rich array of plant and animal life.

This is indeed head-spinningly difficult. It is something that philosophers have been working on for some decades and yet remain in great discord over. I have my own theory, which I think is correct, but I cannot say that many other people think the same. My theory implies that there is a tradeoff between numbers of people and levels of well-being, provided the level of wellbeing is high enough. Adding to the population of the world is in itself a good thing, provided the people who are added live lives that are sufficiently good. If this is so, it means we should be willing to sacrifice some of the level of the well-being of the people there already are in order to have more people with sufficiently good lives. There is a balancing to be done between the well-being of the people who live and the number of people. Both are good things.

Many people think differently. They would like to have just the number of people in the world that would lead to the greatest level of flourishing or the greatest level of wellbeing. They do not count the number of people in itself as valuable. I think there are good arguments in moral philosophy that show this is a mistake, but it is not easy to convince people of that. So, we have a problem in the philosophy of population. Almost every theory, in fact, every theory you can think of, leads to implications that many people find implausible. That is a real difficulty.

A lot of people from the outside looking in on the work of ethical phi-

losophers are going to read us talking about these sorts of ethical dilemmas, about what is the ideal number of humans living on the planet and the kinds of arguments that we are making, and they are going to see something that looks much more technical, that looks much more academic and detached. But these kinds of arguments do filter down into everyday conversation and into policy. This raises a question that I am often concerned with when I read philosophers with interesting theories. How can deeper and clearer thinking about ethics help us to actually become more ethical people? How can deeper and clearer thinking about ethics help to resolve the challenge of global warming?

The main thing is not that individuals need to become more ethical, but we need to have more ethical governments. The governments have to solve this. I think my knowledge of moral philosophy helps me figure out what governments ought to do; it does not help me much with getting them to do it. The only means I have available for moving governments is the one philosophers in general have available. We can work out our arguments as well as we can, and put them into the public domain. Once there they can filter down and inform people's thinking, and people in big numbers can influence governments.

There is some technical stuff involved in moral philosophy. That is not surprising. There is a lot of technical stuff in science, for example, but it does not mean that people cannot take account of science in their thinking. Science can filter through to people's consciousness through the work of people who present it well. That is what we have to do with philosophy. We have to present it as well as we can. We can hope that good philosophy, rather than bad, will influence people's thinking and, in the end, influence the political system.

This is what philosophy has done through the millennia. The work of philosophers over the last two-and-a-half thousand years has gone a long way towards shaping our present world. The influence of Aristotle, for example, is still enormous. The influence of John Stuart Mill, from the nineteenth century, is very big. It has created the liberal system that many of us in the West live under. So, philosophy is extremely influential. But it has to work through the filtering process of people's minds and the political system. I am sorry to say the filtering takes a long time. This makes me anxious about philosophy and climate change, because we have only a short time.

What is the most effective thing we can do to minimize suffering and increase well-being in the world today?

Improving the distribution of wealth around the world would go very far towards improving well-being, because it would solve many other problems as a by-product. Eradicating malaria would be a very good way of doing a huge amount of good in the world. Malaria kills over a million people every year. It sometimes seems it would be possible to eradicate it, and it might be possible to do so with relatively little money. Certainly, controlling malaria would not require the sort of money that is required to make a big difference to climate change. There are other diseases, particularly tuberculosis, that can be relatively cheaply controlled. Providing clean drinking water around the world would be a very effective way of doing good. I think that solving the problem of climate change is not at the top of the list of ways to improve the world. Nevertheless, we should solve the problem of climate change, because climate change will do an immense amount of harm. Solving the problem would be extremely beneficial. Just because it is not the first thing we should do, it does not follow that it is not an essential thing we should do in order to make the world better in the future.

And what is the most effective thing we can do to help slow the rise in CO_2 emissions?

I have a particular thought about that. I can describe it to you, but I am not sure I can justify it in the time we have. We need an international bank of climate change. It needs to be an international institution with a lot of clout, like the World Bank and the IMF, which were set up in the aftermath of the Second World War to help global reconstruction. We need an institution with enough financial power to guarantee loans for investments that will reduce emissions of greenhouse gas.

There are many emissions-reducing investments to be made, all over the world. They will be expensive. We need to switch away from fossil fuel towards renewable energy and nuclear energy. These investments could be financed out of loans if we had a sufficiently powerful international bank to support them. If they were financed that way, they would not demand sacrifice on the part of the current generation. It would not be necessary for the current generation to reduce its own consumption to pay for emissions-reducing investments. Instead, these investments could be paid for by reducing more conventional sorts of investment such as road-building. What is holding up our doing anything effective about climate change is that none of our governments are willing to accept sacrifices on behalf of their populations. When they go to meetings of the United Nations Framework Convention on Climate Change each year, they seem to think they have to divide up sacrifices among the nations – to decide who has to bear the cost of reducing emissions. And our governments refuse to say "Okay, we agree that our population will sacrifice part of its own well-being for the sake of future generations." But in principle we could arrange for the necessary changes to be made without sacrifices by the current generation. If we can do that in practice, then I think we may have a way forward. We might break the log-jam in the international negotiations. This could be achieved by a powerful Bank of Climate Change.

KATE PICKETT

Kate Pickett is the co-author, with Richard Wilkinson, of *The Spirit Level*, which was shortlisted for Research Project of the Year 2009 by the Times Higher Education Supplement and chosen by the New Statesmen as one of the top ten books of the decade. Both David Cameron, Britain's conservative Prime Minister, and Ed Miliband, former leader of the British Labour Party, have referenced its thesis that everyone does better in more equal societies.

Pickett is Professor of Epidemiology at the University of York, a National Institute for Health Research Career Scientist, and Director of The Equality Trust.

THEO HORESH: *You have written, "Economic growth, for so long the great engine of progress, has in rich countries largely finished its work. Not only have measures of well-being and happiness ceased to rise with economic growth, but as affluent societies have grown richer, there have been long-term rises in rates of anxiety, depression, and numerous other social problems. The populations of rich countries have gotten to the end of a long historical journey."*

When did greater affluence stop bringing about an increase in happiness?

KATE PICKETT: It is hard to give an exact answer about the date. What the data show is that for poorer countries and emerging economies, as they get a bit richer, as average standards of living rise, health and well-being improve very rapidly. And then once you get to a certain point of economic growth, you see a leveling off of that relationship. So, among the more affluent countries, there is no longer a relationship between average incomes, or standards of living, and average levels of health and well-being. It is quite clear that in countries like the United States and the United Kingdom, we have had decades of economic growth, obviously not very recently, but in the past, that were not accompanied by any increase in happiness. And although life expectancies for the whole population tend to improve over time, that is no longer a function of countries getting richer and richer. So, the timing will vary a bit from country to country, but it is clear that once you reach a tipping point of material affluence for the whole population, then you no longer benefit in terms of health and well-being from further economic growth.

So, how would you measure the happiness or well-being of a country?

We have been much more interested in objective measures of health and well-being than subjective ones. You could say that the population's health is a good indicator, so things like average life expectancy or infant mortality rates. Those have often been used as indicators of well-being in a country. Economists and social scientists who measure happiness tend to rely on random samples of the population, who are simply asked to subjectively report whether or not they are happy. And you could argue that those more subjective measures are less interesting, but a number of people have been working for quite some time now, in different groups, trying to find a better indicator of well-being for population than gross domestic product per-capita, which is how we measure economic growth. So, we have things like the Human Development Index and indices of well-being developed by different groups. And they tend to include both subjective and objective measures of well being. There are lots of different ways to do it, and you can look carefully at what you think is most important for a population.

You have written that problems that are more common at the bottom of the social ladder are more common in unequal societies. This seems to be the most remarkable breakthrough in your research. Perhaps you could say a bit about how this process works and how it was discovered.

What we find is that many problems are more common at the bottom of the social ladder, but also slightly more common just beneath the top, compared to the very top – they have a social gradient. And this is true of lots of health and social problems. They are more common at the bottom and less frequent at the top, but there is a gradient, rather than a threshold effect. These are not just problems of the poor that the rest of us do not have. These problems are more common when the social ladder is steeper, when the gaps between rich and poor are greater; in a sense, when the social distances between people are larger, when the social hierarchy is more rigid and more fixed. If you take that fact, more problems in more unequal societies, in combination with no longer finding that society's health will be improved as countries get richer and richer, you start to realize that this social gradient is not to do with people having more or less things, more or less material goods.

It is something to do with social ranking itself, something to do with the degree of social stratification in those different societies. The class system would be another way of putting it. And so what we are looking at really is the psycho-social impact of that social status differentiation, that seems to be affecting the whole society in various ways, getting under our skin, into our minds, affecting our emotions, and causing a higher degree of health and social problems in more unequal societies.

Maybe you could say a bit more about some of these health and social problems that show up in the unequal societies.

There is a huge body of research looking at income inequality and health, that very consistently finds links to higher rates of infant mortality, working age mortality, and lower life expectancy. We have also found links to obesity. And increasingly, there are links between income inequality and mental illness of various kinds, including drug and alcohol addiction. So, there are a whole range of health problems. And this is not surprising, because most health problems do have social gradients.

But we also find an effect of income inequality on various measures of what we might call "social cohesion" or "social relationships." The more unequal societies have lower levels of trust, lower levels of social capital, women's status is lower, they have higher levels of violence and more imprisonment, so a whole range of issues that tend to indicate more social dysfunction.

And there is a third area with a strong connection to inequality, which is the kind of thing that economists call "human capital development," the educational achievement of young people, their tendency to continue in education or training after schooling age, rates of teenage pregnancy and birth, and measures of social mobility as well. So, those are the big three areas in which we find a significant influence of income inequality.

Perhaps you can say a bit as well about the statistical methods you used to attain these results.

What we presented in our book, *The Spirit Level*, were quite simple statistical analyses correlating measures of income inequality taken from the World Bank, and when we are looking at U.S. states, from the U.S. census, with health and social problem outcome data, generally drawn from the most reputable international sources we could find - so health data from the World Health Organization, vital statistics data from different counties in the U.S., educational attendance in the programs for international student assessment, etc. And we present very simple correlations in our book.

But actually, there are many statistically much more sophisticated analyses of these same issues. Some of them are what we call multi-level studies, which look at the impact of income inequality after controlling for people's individual income and education. Some of them are time series analyses looking at changes in income equality over time and changes in outcome. So, there are a range of different methods. And we are really seeing a burgeoning of this area of research and a consistent picture coming through whatever kinds of methods are used.

Going back to how inequality affects our self esteem, Ralph Waldo Emerson has written, "'Tis very certain that each man carries in his eye the exact indication of his rank in the immense scale of men and we are always learning to read it." How does this experience of reading social rank impact our self esteem?

It is quite clear from psychological work and experiments that this is very true, that we do experience ourselves through other people's eyes, and we know where we stand in relationship to other people. Amazingly, psychologists tell us that we judge each other's social rank within the first few seconds of meeting. So, we are obviously doing it on a very subconscious level in response to lots of subtle clues, including the way we dress, the way we hold ourselves, the way we speak, our demeanor. How does this affect our self-esteem? Well, it depends on our social rank. If you are extremely highly placed in society and confident of your place there, then seeing yourself through the eyes of others must feel very reassuring much of the time, but for everybody else, this is a constant source of stress.

Indeed, there was a meta-analysis of over 200 psychological experiments, where they were trying to understand what kinds of things cause the most stress. They were putting people into experimental situations, trying to stress them, and measuring the stress-hormone response. And they found that the most consistent stressor, the thing that was most likely to give a stress response, were situations that contained what they called "social-evaluative threat," those situations where other people have the chance to judge us and judge us negatively. It is clear that this reading of social rank has an effect on how we feel, and indeed, has an effect on our physiology, because it is certainly not good to have your stress hormone levels chronically raised. Obviously, this is a source of chronic stress for an awful lot of people. And it is quite clear that this is a mechanism through which greater inequality, a steeper social ladder, and more concern in most societies about the meaning of rank, has an effect on other individuals.

As I was reading The Spirit Level through for the third time this year, I began to wonder what it would be like, how my life would be different, if I had been brought up in a much more equal society, how I would have approached my life. Can you talk about the difference between living in an equal society and an unequal society from the inside?

Whenever I am in, for instance, the Netherlands or in Sweden, I do seem to relax almost immediately. And it is hard to know how those things all operate, but one thing we often ask people to imagine is what it feels like to live in a country where most people trust each other, compared to what it feels like to live in a country where most people do not trust each other. There are huge differences in trust in different countries, strongly related to income inequality. In the more equal countries, like the Scandinavian ones, about two-thirds of the population feel that everyone can be trusted. And it drops to less than one-fifth in the most unequal countries. And you start thinking, what does that mean for social relationships? What does that mean if you are a woman walking home alone at night, or what does it mean on the school playground, what does it mean in the workplace? I think you could start then to imagine some of the tensions and some of the anxieties that increase with that lack of social cohesion and are simply less salient in a more equal society. Feeling free of that anxiety about rank and status, I think, worries and drives us all a lot of the time.

So, how does inequality affect our ability to cooperate with others?

Again, the data on trust are very illuminating. It is extremely difficult to cooperate with people if you can not trust them. In a way, we have got two strategies available to us as human beings when we are in relationship with one another: we can compete for things or we can collaborate. We can obviously do both quite effectively. As human beings, we have lived in extremely unequal and very competitive societies, and we have lived in much more egalitarian and cooperative ones. But clearly, if rank and status and inequality are inhibiting social cohesion and social trust, and increasing the distances between us, then that is a barrier to cooperation.

America tends to score worse than almost every other rich country on almost every indicator of social development. We have the highest murder rate, the most prisoners, and the highest level of mental illness, teen pregnancy, and infant mortality. We have some of the lowest levels of social trust, some of the highest levels of drug abuse, some of the lowest levels of childhood well-being, and shorter lives than people in most other developed countries. Even though we are one of the few richest countries, we come close to giving the least in foreign aid, recycle less than any country other than the UK, and have the biggest ecological footprint after the United Arab Emirates. Why are these environmental concerns also correlated with social and health concerns?

A couple of reasons, really: one is what we were talking about before, about cohesion versus competition. Do people in society act more individualistically or do they act more collectively? Are they thinking about the common good, or are they thinking about themselves. That is part of the explanation why we see more giving in foreign aid and more recycling and more attention to environmental concerns in more equal countries. Indeed, we have found that even business leaders in more equal societies are more likely to say they think their government should comply with environmental regulations. So, there is a sense in which more equal societies, as well as being more cohesive and being better able to act collectively in ways that are better for their own society and for the environment, are also more willing to reach out to other countries. They score higher on the Global Peace Index as well, for example. Those are important ways in which the ability of a society to do what is right for the environment, in a sense, mirrors their ability to cooperate.

Reducing greenhouse gas emissions is going to require individuals, businesses, universities, and governments to make changes, but it often seems we are too caught up in our own personal and institutional worries to make these changes. How might reducing economic inequality help us to bring about the kind of changes necessary to halt global warming?

There is a very important message here and it is an optimistic one. If crime rates come down, if health is better, if kids are doing better in school, if social relationships are better, then the quality of life in the whole society improves. That is quite a plus that you can offer people in exchange for perhaps having to give up consumerism, driven by status, driven by a need to make more money. You are replacing people's drive towards accumulating more material goods, which we cannot afford as a planet, with a higher quality of life. And actually, if you ask people what is most important to them in life, almost none of them will say my new car or my new handbag or my bigger house or the time I spend at the office. Almost everybody would say that the thing they value the most is the time they spend with family and friends.

It is the quality of our social relationships that matters more to us than anything else. And we can say to people that at a lower average standard of living, at a lower level of gross domestic product per capita, you will not sacrifice anything in terms or health. You will not sacrifice anything in terms of happiness. And in fact, you will probably have a reduction in lots of different health and social problems, that is a real plus to help people get engaged in thinking about what we want outsides to look like. What is the economy for? The economy should be there to serve us, and we need to make that work in balance with the environment.

You have hit on a very big lever. It is amazing how many progressive concerns can be addressed through simply addressing income inequality. But then you also point out in your book that this does not have to be done through government programs. You point to Japan, which does not have high levels of taxation, as being a very equal society because of its greater equality of income. And you point to the state of New Hampshire and the state of Vermont, both having some of the highest levels of economic equality in the U.S. But while Vermont's has been brought about through redistributive taxes, New Hampshire achieved the same thing, right next door, through greater income equality, largely because of the presence of strong unions. Perhaps you can say a bit about the many options for implementing greater economic equality.

Another very optimistic aspect of what the research shows is that it does not seem to matter how you reach your level of greater equality, whether it is through redistribution in a country like Sweden, or whether it is through having smaller income differences to start with, like Japan. What that means is that there is a whole suite of possible policy levers for changing the level of equality in society, some of which are going to appeal to people at one end of the political spectrum, and others might appeal at the other end. But really, there is something for everybody. So, if you are interested in redistribution, you can obviously impose a more progressive tax regime. There is a lot that could be done with tax havens and tax avoidance to increase the funds available for redistribution - certainly that is an important part of what can be done. But the problem with doing it all through government implemented strategies is that if you have a change of government, those can be changed as well. So, that makes your level of equality or inequality a little bit vulnerable to changing political climates.

We are more interested in seeing greater equality come about through a rise in economic democracy of any kind, really working towards small income differences to start with. And there are lots of different ways in which that can be achieved; anything from appointing employee representatives onto company boards, and particularly on those boards that set the remuneration level for chief executives and the senior management staff; stronger trade unions have been clearly beneficial for keeping income differences smaller, and we do see stronger trade unions in more equal societies; but also forms of employee ownership of companies, employee share-ownership, cooperatives, mutual societies. In all of those forms of business or institutions that are more democratic, you tend to see smaller income differences.

You have also pointed out that less equal societies are less innovative. Whereas the highly unequal U.S. and Singapore have some of the lowest numbers of patents per million people, highly egalitarian Finland and Sweden have some of the most - this really surprised me. Would greater economic equality also inspire more clean energy innovations, and if so why?

It seems likely, and it is quite a surprising finding. A lot of people seem to assume that you need a certain degree of inequality to drive competition, to drive innovation, that it is only when you have inequality that people will be ambitious and creative. This might work for a few people, but really the major impact of greater inequality is the wastage of human potential and human talent. So, if in a more unequal society levels of child well-being are lower, which we definitely see, if educational attainment is lower, if more kids drop out of high school and fewer stay in education and training, then you are wasting an incredible amount of human potential and talent. And in more equal societies where a higher proportion of young people do better, then maybe that is the environment in which they can be more creative and innovative.

What is particularly sad for the United States, in particular, is that its social mobility is much the lowest of the countries we looked at. So, it is much harder for people to escape their class of origin, to forge a life trajectory for themselves that is not dependent on what their parents did or achieved. And that is in complete contrast to America's ethos, to its sense of itself as a land of opportunity. It is in complete contrast to the idea of the American dream. It is a lot easier to achieve the American dream if you live in Finland or somewhere else in Scandinavia. And I think it is very sad that a country that was founded on principles of freedom and opportunity actually turns out to be the one where your life trajectory, your life chances, are most dominated by the chance of birth, by the accident of birth.

And inequality really affects the way we think of and plan for the future as well.

If you live in a very unequal society, the way you might think about your own life chances is probably very different to how you might think about them if you live in a more equal society. There is a closing off of opportunities for certain people in more unequal societies. It seems clear that you cannot really have equality of opportunity, which everybody is in favor of, without more equality of outcome. And it does affect us as individuals. As societies, again, it does affect the way we view our public policies. It affects the way we think about who the economy is for, and for too long, in the more unequal of the Western developed countries, like the U.S. and U.K. at the moment, our political and public policies have really been driven by the interests of the super-wealthy and those who are already very well off already, rather than thinking about planning for the future and planning for distributing the benefits of the economy more broadly across society.

And according to your analysis, economic inequality also drives consumerism.

In a more unequal society, status matters more, so that is the key, because we express our status in modern societies through our ownership of various things. We tend to express our status in the clothes we wear, the cars we drive, the size of our house, where we go on vacation, what jewelry we wear, what food we eat. All of our purchases are designed to reflect our status. That is really a zero-sum game, because it just keeps ratcheting up and getting harder for people to keep up. And so if the key driver of consumerism is the need for status, and the key driver of the importance of status is economic inequality, then obviously, if you want to reign in consumerism to help face the challenges of climate change and move towards more sustainable economies, then dealing with inequality is key.

Now one of your more striking arguments is the idea that we are not going to be able to save ourselves with new green technologies in the unequal societies, because in the unequal societies, this urge to consume, to play this status game, to do the conspicuous consumption, and to reaffirm ourselves by demonstrating, through the objects we own, that we are okay, will fuel a cycle in which we find other places to spend the income that is saved on energy. This is a really bleak prospect.

It is, because if you do not remove that driver, there is no incentive for people to stop. So, if we introduce wonderful energy saving devices that cost less, and I am not allowed to buy a gas guzzling, huge car anymore, then I cannot express myself through gas guzzling, huge cars. And so I am going to spend my money on something else that will do it. And anything else I consume is obviously going to consume the world's resources as well. So, it becomes harder and harder to think of ways that you can actually create a more sustainable economy, unless you remove people's drive to have more stuff of any kind.

You have quoted Henry Wallach, former head of the Federal Reserve saying, "Growth is a substitute for equality of income. So, long as there is growth, there is hope."

But one of the greatest hopes of environmentalists is that we might shift to a more qualitative sort of economic growth, what John Stuart Mill first described as "a steady-state economy." How does this hope match the realities of more equal societies, like Denmark, Sweden, and Belgium? Are they closer to this ideal of the steady-state in which the increase in material usages is halted and economic growth begins to show up in higher quality services, innovation, and creative work?

They are certainly closer than the more unequal societies, but I think even those countries would recognize that they have quite a way to go. And sadly, Sweden has been becoming quite rapidly more unequal than it used to be in the recent past. So, there are many, many challenges there as well. I do not think we are really seeing any of the large nations starting to grapple with what it means to think about a steady-state economy, what it means to create a more sustainable economy yet. I do not hear politicians, really in any country, yet having the courage to say that they will not pursue economic growth. They are unwilling to give up that mantra, even though they know that the benefits of economic growth do not trickle down to benefit most of the population, even though they know that we are coming up against the limits of natural resources. So, we are lacking leadership with real political courage.

And because coping with the challenges of global climate change is going to require international cooperation and international agreement, we need to find ways in which we can get countries to work together. They are so unwilling at the moment to give up any perceived competitive advantage, to give up any possibility of maximizing their own economic growth. And until we can find a way to make that work, and get the rich, developed countries together to say they are going to reign in growth, whilst allowing the developing, emerging economies to continue to grow, then I think we are in for some rough times.

Your research focuses on the inequities within rich countries, and within the American states, but it is often argued that if we are to halt the rise in global warming pollution, stop the destruction of rainforests, and reverse the decline in global fisheries, we need to be able to think globally and to expand our circles of ethical concern to include the whole of the world. But economic inequality within the whole of the world is far worse than inequality within rich states. How might we bring about greater global equality, and is greater global equality necessary to bring about transformation on global challenges, like climate change?

That is a big question. It is quite true that economic inequality in the whole world is very large, and of course, within many developing countries, income inequality is much, much greater than anything we see in the western developed nations. We do need to think about how we can work to bring about greater global equality, because I do think it is going to be essential if we are going to get societies together, communicating together and cooperating, but also if we want to prevent mass migration and conflict in a world where too many people have far, far too little, and a few people have very much.

There are lots of reasons why greater global equality is necessary for the sort of international cooperation we need. How to achieve that? It is quite hard to say. It is obviously going to require changes in the way countries tax, and international agreements to avoid multi-national companies, or small oligarchies, using other countries to avoid paying tax in their own. We do need international action on tax havens. If we had international action on arms sales that would be helpful as well. If we could reduce conflict and avoid having to see so much spent throughout the world on militarization, then there would be lots more money available to redistribute to make societies more just and more equal. Perhaps we could put in place something like an international social-security program. There are lots of different mechanisms.

But I think the first step towards that, which is obviously very difficult, is to get a wider public and political recognition of the impact of inequality. Far too many people still do not know about the impact of inequality. They know about poverty but they do not know about inequality. So, we do need to educate the public, and educate politicians, and get them to see the benefits of greater equality before we can hope to work towards that sort of change.

We live in an era of increasing economic and social globalization. As humanity becomes more globally conscious, do we need to worry about some of the same social and health problems we have found within unequal nations also showing up between nations in an unequal world?

We are already there in very exaggerated forms, I think. We see much, much bigger differences between nations than we do within them. We have whole countries where life expectancy is extremely low, countries where infant mortality is very, very high. So, we already live in a very unequal world. And people who live in countries with very high levels of health and social problems already are no longer isolated from the rest of the world. They are connected, they are aware. And the proportion of people living in Africa, who have access to the Internet through a telephone now, is astonishingly high. And as we start to become aware of the lifestyles of some, which are really taking place at the expense of the lifestyles of the vast majority, that is a recipe for unrest. It is a recipe for mass movements of people, economic migrants seeking the best in life elsewhere, and extremely high potential for conflict and problems. So, I think there are a lot of good reasons to bring about greater equality for social justice, but also for peace.

We have covered a lot of ground in these questions. Part of the reason for this seems to be that your research just touches on so much. Perhaps you can wrap up with a summary of all of the ways that greater economic equality might be able to help us in slowing the rise in greenhouse-gas emissions and halting global warming.

Greater equality improves social relationships, levels of social cohesion and trust. That means that the populations of more equal societies are more likely to be able to act together for the greater good and to be more interested in the public good and less individualistic. We need to get people's agreement to think about changing their ways of life, and so that improved social cohesion in more equal societies is key.

Greater equality also reduces economic instability. So, in more equal countries we see less boom and bust, more stable economies, and situations that allow populations and governments to actually plan for the future. When you are in a situation of instability, it is very hard to think about doing anything other than getting yourself out of it. Inequality drives status competition, which drives personal debt and consumerism. In more equal societies we also see more creativity and more innovations that are more likely to perhaps get some of the key ideas we need for creating fantastic new green-technologies to help us address climate change. So, I think in all those ways you can think about greater equality having a key role to play in creating more sustainable economies, and coping with the challenges of climate change.

ROBERT HENSON

Robert Henson is a science journalist and meteorologist. His Rough Guide to Climate Change was shortlisted for the UK Royal Society Prize for Science Books in 2007. Newly updated for its fourth edition as *The Thinking Person's Guide to Climate Change*, the book has proven itself to be one of the more lasting and comprehensive introductions to the topic. Henson is also the author of *Weather on the Air: A History of Broadcast Meteorology* and has worked with the National Center for Atmospheric Research and its parent organization, the University Corporation for Atmospheric Research, since 1989. He has won several Distinguished Technical Communication awards in international competitions sponsored by the Society for Technical Communication and has also has published widely in *Nature, Scientific American, Discover, Audubon, Sierra*, and dozens of other forums. In 2005, Henson began covering weather and climate science as a meteorologist and blogger for *Weather Underground.*

Combining scientific rigor with a talent for making the complex simple, and a curiosity for the yet-unknown, Henson demonstrates yet another way to think and feel about climate change—one that draws on the inherent fascination of weather, so intimately tied to the way we live. The curiosity that Henson brings to his writings provides a gateway for skeptics and a source of renewal for weary veterans of climate wars. Henson's passion for weather is also evidenced by his many years of sky and storm photography and his participation in thunderstorm research. He lives in Boulder County, Colorado, where he is an avid cyclist, vegetarian cook, and cinephile.

THEO HORESH: *Climate scientists are in many ways besieged. Before them, they have this great challenge to civilization whose outcome is very difficult to predict. And while they have been empowered to study climate change, many in the scientific establishment would like to limit their ability to advocate on the issue. Meanwhile, climate skeptics are well-funded, often highly aggressive, and they have the inertia of personal and institutional habits on their side. So, climate scientists are in a pickle. They understand the problem better than anyone else, and yet they are limited in their ability to tell us what to do about it. How do you see most climate scientists responding to this bind, and how do the climate scientists you know tend to do advocacy within these constraints?*

ROBERT HENSON: Boy, it is a pickle, isn't it? There are so many different ways to approach it if you are a climate scientist. I am thinking right now of one good friend of mine, who is a researcher into aspects of drought and water excess. He wrestles with this a lot. He cares deeply and passionately about the subject, he has spoken to many groups about it, and at the same time, he is not an advocate in terms of saying, "We need to do X, Y and Z." That is where a lot of climate scientists fall out. Their focus is on the science, and rightly so. They are trying to figure out what is going on with this amazing, complicated system we call the Earth system, which of course includes the atmosphere, the biological components of the planet, the land, and the sea ice.

It is appropriate that the climate scientists focus on that. But of course, there are mammoth implications for climate change, and you could easily say, "Shouldn't the scientists be out telling us how we need to fix this?" And many have decided that it is really not their place. There was one researcher involved with one of the IPCC reports who literally said that on the radio in an interview, that it was not her job to tell society what we should do about this. So, it is a pickle all right, as you aptly put it.

One sort of middle way that can work really well is where scientists are not prescriptive, but descriptive, in terms of policy. And that often involves something like this: you map out what you expect could happen to the climate in the next X decades, looking at whatever piece of climate you are an expert on, and then just point out that if emissions do not change, we are on this path. If we reduce emissions, then we are much less likely to have this outcome. And it is really up to society to decide how important this is relative to other things. So, you are not telling people what to do, but you are making it clear that if you take one path you may get a different outcome than if you take another path.

It is almost like a therapist giving a patient ideas about what could result if they act in certain ways. The therapist might have a preferred outcome, but a skilled therapist is not going to say which and believes, in fact, that the patient should have and has agency into what the outcome should be. It may sound patronizing to say a scientist is like a psychotherapist, and the world that responds to climate change is like a patient, but climate scientists have sometimes been called "planet doctors." So, maybe it is not such a stretch. The part that appeals to me personally, though, is to be as descriptive as possible and just lay it all out for people. And by and large, I like to think society eventually acts in a smart way.

It often looks like the environmental advocates go too far in portraying climate science as heralding the end of the planet, not just the end of the planet as we know it, but literally the end of the planet. Even when they do not go out and say that, it is easy to come away with this impression. And then on the other side, we have a type of climate skeptic who does not necessarily deny that climate change is happening, or even that humans are causing it, but who is skeptical that the results will be very disruptive at all. And it must be difficult as a climate scientist to watch these extremes play out when you can bring so much more balance and depth to the discussion.

The IPCC reports do a masterful job of pulling the science together in a very sane and coherent way that, unfortunately, has some unavoidable jargon in it - so it is not for everybody. But they really do a fantastic job of laying out the implications and the science in a way that makes it clearer we are changing the planet and we could change it in profound ways. And yet it does not mean the end of the planet. There are not a lot of exclamation points in the IPCC report; in fact, I do not think there are any. It is not grabbing you by the shoulders and shaking you. And some would argue, and I think there is validity in this, that people need to be shaken up and need to realize that this is a serious matter.

But you cannot scare people into acting. There is a fair amount of social science research on that. If you frighten people too much about climate change, it can make them just turn off, because it is such an abstract concept and the results are so decoupled from any one action, from starting your car or taking a flight. You are not going to see an immediate response to that in terms of climate. It is easy for people to turn off, consciously or unconsciously, to the problem. You have to almost entice people to think about it and take it seriously, but not end-of-the-world seriously. I think that is where hardcore environmentalists sometimes boss us around a little bit. Obviously, there are deep beliefs in the whole idea of the end of the world, in religion and other aspects of culture. So, it can be easy to fall into that groove and to see this as the way that the world will end - "We all know the world is going to end somehow and this is it." Clearly, we will have a profoundly different climate in a hundred or three hundred years, and yet it will not be the end of the world. Personally, I do not even think it will be the end of societies, but I do think there will be huge implications. Cities like New Orleans and Miami are going to have to be literally transformed or they will not be there as we know them when sea level gets beyond a certain point. They are not going to look the same by any means. It will be a different world in a lot of ways, but I still think many aspects will be recognizable, and I am not one for dystopia.

It astounds me how almost every look at the future, in terms of entertainment, is a dark look. It is almost like we are hard-wired to think the future is going to be bad. There is all kinds of evidence to the contrary. Steven Pinker has written a book on how people are generally nicer to each other now than they have been throughout history, and I am a big believer in progress. Yet, I am also a believer in the dangers of technology, and I do not think it is invariably good, nor that we can use it to get us out of this one. It is going to take some hard choices and some ways of rethinking how we live and how countries interact with each other. But I am an inveterate optimist.

One of the unique things about climate science, in relation to other scientific disciplines, is that it is dealing with so many indeterminacies. As the Arctic sea ice melts, the planet will get darker around the North Pole and the Arctic will absorb still more heat. How much, we do not know. As temperatures rise, forest fires and disease will kill off some portion of the planet's trees, and these will release still more CO2, further heating the planet. How much, again, we do not know. We do not know how humans will respond to the climate changes that will happen over the course of say, the next half-century, so we do not know how much CO_2 emissions will be emitted into the atmosphere. And even if we did know, we would not know how the flora and fauna of the world would respond, because there are just so many indeterminate feedback loops involved. The list of unknowns goes on and on and on. What is it like to do science that is so indeterminate, and what is it like communicating such an indeterminate science?

It is quite a challenge. You have so many variables to juggle, and they all influence each other. When they first came up with climate models that depicted not only the atmosphere but the ocean, they were called "coupled climate models." I sometimes wonder if that was because they figured that was the only coupling that would ever happen. In that case, you can use a term that implies just two partners. But now you have models that incorporate the biosphere, land-model interactions, the ice system, or cryosphere, and so forth. But they are still called coupled climate models, even though they are really polyamorous climate models. So, it is complicated, there is no getting around that.

When people say the debate on climate change is over, what it is that can now be said to be settled science?

That phrase always makes me a little nervous, because science by its nature is never completely settled. Gravity might stop working some day. We do not think it will, and there is immeasurable evidence that it will not. But that is never an impossibility. So, science is always, in a sense, tentative. Obviously, we live our lives counting on science to work, and by and large it does. But it is largely settled that carbon dioxide and greenhouse gases, in general, warm the planet, and as we put more of them in the atmosphere, it is going to cause our atmosphere and the oceans to warm. In fact, most of that heating has gone into the oceans, and that is something that is not well known by the public and a really important piece of the puzzle.

So, when people say that the climate models do not work, that they do an inadequate job of modeling past and potential future changes in climate, it seems this really does not matter much. You could take away the climate models and you would still have this very basic understanding that greenhouse gases released into the atmosphere warm the planet, and that would be enough to take this problem very seriously.

It makes me think of the analogous situation with health. We have known for decades, if not hundreds of years, that if you walk every day, eat your vegetables, and get enough sleep, you are more likely to enjoy good health. Modern medical science continually re-proves again and again that those are core fundamental aspects of staying healthy. Likewise, there are aspects of the climate system that we have known for decades, perennial truths we have proven again and again.

One of them is that when you increase greenhouse gases, the atmosphere warms. The planet has been warming for a hundred years and we have increased greenhouse gases for that entire period of time. Since we began measuring carbon dioxide in a reliable way in the 1950s, the concentration in the atmosphere has increased roughly 35 to 40 percent, and the atmosphere has warmed by something between a half degree and a degree Centigrade, so a little over 1 degree Fahrenheit. That warming is actually over a century, but most of it has been since the 1950s. And those changes are in line with what modern models tell us will happen over decades; that pace of warming is roughly on par with what the models tell us. In fact, if you go all the way back to around 1900, Svante Arrhenius, a researcher who won the Nobel Prize for some unrelated work, actually calculated the warming that would happen with a doubling of carbon dioxide, and he was not too far off the mark. So, I would call that settled science. The understanding that sea level will rise is also pretty settled - it is happening now. There is no reason to think it will stop.

The idea of a sped-up hydrologic cycle is probably not in the same realm as the others, but it is pretty well understood and accepted. There is more moisture in the atmosphere through a well-known physical relationship called the Clausius-Clapeyron equation. As the atmosphere warms, you get more water entering the atmosphere, and this causes it to rain and storm more heavily in places where there is frequent rain and snow, but it dries out more in drought-stricken areas. That is a pretty well-understood system, but there will probably be nuances refined as time goes by.

Things are less settled in the realm of specific weather events and types of weather, things like hurricanes, severe thunderstorms, and tornadoes. There is still a lot of active research on what is going to happen with those. And different studies will come along and threaten to upset the apple cart, and then other studies will negate those. As soon as an understanding starts to coalesce, it gets rattled a little bit. The basic idea of hurricanes gaining in strength, that the strongest hurricanes will get stronger, has grown in acceptance over time. But there is still some controversy over what will happen to the number of hurricanes.

Now, each of us imagines something different when we are told that temperatures are likely to rise say, 5 degrees Fahrenheit over a given period of time. Some of us are going to visualize Nigerian villages, and we are going to wonder how the people in those villages will fare. Some of us will concern ourselves with low-lying cities in places like Bangladesh. Others worry for the fate of the rainforests and other species. And still others turn their minds to the American generation of say, 2100. The point is that we all have different concerns, and when we think about climate changing and how it plays out across space and time, we are each going to think about different things. So, what tends to worry climate scientists the most?

There are quite a variegated number of climate scientists in different areas, and their concerns largely reflect some of the dimensions you are talking about. There are some who specialize in sea ice, for example, and I know several of them, who are profoundly worried about the fact that some day soon we may not have sea ice for at least part of the summertime in the Arctic. Certainly, that seems very likely by the end of the century and possible even within the next couple of decades. That could deeply affect polar bears and all sorts of wildlife in that part of the world. Folks who specialize in agriculture are very concerned about how that could evolve, especially in the tropics, where you do not really need a lot of warming to go beyond the bounds of what we have known as climate there. The temperature swings are generally larger in the poles, and the amount of warming expected there is greater, but they are more used to variability up there. Temperature variation through the year, and even through the day, tends to be less in the tropics. So, a little bit of warming goes further in terms of changing the climate to something less recognizable, and it could have big impacts on agriculture.

So, I tend to think that climate scientists who specialize are naturally concerned about those areas in which they specialize. But there is obviously also a shared concern about the surprises that could come along. One of the biggest concerns about which several climate scientists are now worried is the things that the models may not yet capture. That is not to say that the models are holding out hope that we will not have climate change, but it may manifest in some different ways.

As I am listening to you roll out all of these different concerns that climate scientists are likely to have, it makes me think that climate change is like a thief in the night. But it is a thief that strikes numerous locations at once. He does not just rob a bunch of different banks. He robs some houses, robs some offices, steals some cars. And you just do not know where he is going to strike next. As a communicator on climate change, how do you draw all of this together into a narrative that is going to move people?

That is a really interesting analogy. It seems to be not only one thief in the night, but a thief who has a whole squadron of comrades, who can pose as other people. So, you are not really quite sure what is making mischief in a particular area. One of the perennial difficulties lies in looking at where we will see climate change. Everybody wants to know what will happen when we finally see a weather event and will be able to say, "*That* is climate change" and we see the fingerprints. And I just do not think that is possible in most cases, because there will always be weather. There is natural variability in the system. At some point, the signal rises without the noise, but even then the signal is still going to be expressed in terms of weather. It is not like anyone can say, "Boy, we had some rough climate today," because it is always going to be weather.

Take a storm, like Hurricane Sandy. There are some big questions about the path it took, which was extremely unusual - virtually unprecedented in records that we have going back more than a hundred years. That path may or may not be related to a changing climate. But we do know that the sea level affected the storm surge. And interestingly, there is a different amount of sea level rise going on in different parts of the world because of how ocean circulation interacts with atmospheric circulation. So, the northeast coast of North America is actually getting more sea level rise than some other coastlines around the world. The New York area has had almost a foot of sea level rise in the last century. You can legitimately say that the topmost foot of water in Sandy, that topmost foot of flooding, could be attributed to climate change, because if the last hundred years had not had that warming, you would not have had that extra foot or so of flooding if an identical storm to Sandy came along. Now, it is possible that Sandy would not have happened that same way; you can never completely deconstruct it in a sense. But I think it is one of the few areas where you can make at least a stab at saying what piece is climate change and what piece is probably weather. There are not many of those.

There is a whole interesting cottage industry of research in what is called *detection and attribution*, which is where you simulate a weather event in a computer model, and simulate the atmosphere that led to it over and over and over again to see how many times that particular outcome happens. That gives you some idea of the odds of this event happening. You do this both for the current atmosphere and then for an atmosphere without greenhouse gases, and by comparing the two you can say to some extent how much the odds have been boosted by the presence of greenhouse gases. That is still fairly new research, but it is evolving quickly. And I know folks involved with this who hope to develop it to where they can say, maybe within a few days of an event or even sooner, "Here is how much climate change had to do with this event."

Still, the weather events are going to happen, and we will not really know whether that thief in the night was climate change or not, or how much of it was climate change. But we know that climate change is there in the mix. There is no getting around that. It will continue to be in the mix, and its influences in various areas should only increase over time. That is a really important thing to remember. It is not like we are going to go from now to some end state, but it is a process that will continue through 2050, through 2100. We are in this for the long haul, and it is not going to stand still. We are not going to get to some new climate that we adapt to. We have got to continually adapt to this continually changing climate, which is changing in both subtle and not-so-subtle ways.

Many dismiss the results of climate science because it is such a young discipline. But the early climate scientists stretch into the nineteenth century. Maybe you could say a little bit about how climate science came into being, and how it might evolve over the course of the next half century or so.

The roots of climate science are in hard-core physical science, such as physics and chemistry and to some extent biology. Understanding of the chemistry of the atmosphere goes all the way back to the nineteenth century when we first realized what greenhouse gases were. Some really signal accomplishments in understanding the physics of the atmosphere, like frontal theory, came about in the first few decades of the twentieth century. The real catalyst that set the world of climate science on fire, though, was probably the advent of computers, because before then you just did not have any hope of simulating all the interactions in this amazing atmosphere. In fact, there were people in the nineteen-fifties and sixties who considered climatology kind of the backwater of atmospheric science. Climatologists were the dweebs in the back room, who would keep track of what the highs and lows were. So, not to put too fine a point on it, it was not the most exciting aspect of atmospheric science for a period.

And then by the time you got to the sixties, when computation came along, people increasingly recognized, especially with things like the Mauno Loa carbon dioxide observations, that we could affect the atmosphere. Some people got a bit wild-eyed with the idea, during the Cold War, that we might be able to control the atmosphere to our liking. And while it became clear that we probably could not do that, there was this growing realization in the fifties and sixties that we might actually have a bigger role in the atmosphere than we thought. Even in the Lyndon Johnson administration, there was a report about the potential of climate change. Interestingly, in the seventies there was the recognition that the Northern Hemisphere had been flat temperature-wise since the forties and had actually cooled a little bit. There was a lot of publicity about the potential ice age and snow blitz that could swallow up the northern hemisphere in snow pack. This made major coverage in *Time* and *Newsweek*, but not every climate scientist was going that direction. Many were more concerned about carbon dioxide even then. In fact, the first article in a major journal to use the phrase "global warming" was in the 1970s, in the middle of this period.

So, interest in greenhouse gases never really completely went away. It just took a back seat for awhile. And then in the eighties, when the globe really started warming and when the effects of greenhouse gases simply rose through the floor, the discipline really took off, both in the public mind and in funding. And of course, with additional support climate science could evolve and become more sophisticated and carry out more complicated experiments. So, to some extent it has gone hand in hand with support and public interest, but the roots go way, way back.

As you know, the development of computing power was often spurred on by the need to understand weather better. And few people seem aware that climate science was often at the cutting edge of the development of computing power. There was just so much that needed to be included just to model weather patterns. But it seems that to really model climate change we have got to consider everything that is happening in the world. We have to consider patterns of weather, ocean circulation, and geology. We have to consider ecosystems. And we have to consider human institutions and behavior. The list just goes on and on, and I am curious where it is going. How are these models related to the computing power that we have available? How much can we include in the climate models that are being developed? And does understanding climate ultimately mean understanding the physical world as a whole?

There probably is a limit to how far global computer models can go. You cannot literally model every square inch of the world. It is going to be years and years if we ever had the computing power to do that.

That reminds me of the Jorge Louis Borges story, where they create a map that is so detailed that it has to be placed over the territory, and eventually the map just fades away and covers the territory.

Which brings to mind the idea that no matter how hard you analyze this, there is that much detail. There is still such a long way to go. On the one hand, some of the most powerful computers in the world are being put to this problem. But on the other hand, a lot of climate scientists are keenly aware how much more they could do if only they had more time, because there are only a few computers that are dedicated to the problem. Part of the problem is that they are not bigger, and part of the problem is that there are not enough of them. If there were simply more of those large machines so much more science could be done. But it is expensive, and there is only so much funding.

There are many ways to present information, and that is sometimes a neglected area, because climate scientists are rightly focused on literally getting the modeling done and analyzing the results. But in terms of explaining it to the public and policymakers and stakeholders, there is tremendous work to be done. Climate model interpretation and climate science interpretation is going to be a really critical area, because people ultimately need to act on the information. One might ask, if you get a study and no one responds to it, does it make a sound in the forest?

Then there are the surprises. A good example of that is what is known as the *hiatus*, the lack of atmospheric warming in a dramatic way since about the late nineties. It stayed very warm—the decade of 2000s was actually warmer than any for the last several hundred years, as far as we can tell—but global temperatures were not hitting new record highs regularly. Depending on which data set you look at, there might have been one or two in the 2000s, but generally it has been a very warm but flat period.

It turns out that the heat was basically being stored in the oceans at a more rapid rate than it had been before. We still have global warming, it is just manifesting in the water. Eventually that heat will be released to the atmosphere one way or the other, and the atmosphere will continue to accumulate heat. So, we will see the temperature rise again. It is just a matter of when, not if. But every year that goes by without that happening has provided more opportunity for skeptics to come along and foment uncertainty and a say, "Look, global warming stopped in 1998." It is a very appealing line, and in a very common-sense way someone might say, "Well you know the whole idea of global warming is theoretical, because the temperature is very flat."

Still, the hiatus is a really interesting question, not so much that we do not know whether the heat is being stored in the oceans, but it is not clear exactly how and on what time scale that process happens. What makes the ocean start absorbing heat? What makes it release heat? It seems to be partially tied up with changes in the Pacific. It always seems to tie back there - all roads lead to the Pacific. There is a phenomenon called the *interdecadal pacific oscillation or the pacific decadal oscillation*, depending on what part of the ocean you examine. These are modes in which the Pacific takes up heat over a few years and then releases it over a period of years or decades. That is an example of an area of science that was not being studied really intently until circumstances forced the matter. Part of the future of climate science is responding to how the planet works its way through this warming era we are coming up on.

Now I think your Guide to Climate Change is the best introduction to this material that I have come across, and what I like so much is the way you gently introduce the reader to the science with a wide array of examples, keeping it fresh and interesting in a non-textbook like fashion, bringing all of your curiosities to bare on the material, with a balanced tone. And this sense of curiosity and openness to the challenges to climate science, often from climate skeptics, seems simply an example of good science - it seems to be what scientists are supposed to do. But such a light-handed approach can also downplay the seriousness of what is really a monumental challenge. What is it that leads you to approach this subject of climate change in the way you do, both in your book and in this interview, and how might more of the kind of curiosity you exhibit change the way we all think about climate change?

My intent all along was to reach what I originally thought of as the middle ground in what I was thinking of as a very polarized debate: not to preach to the choir, not to argue with the contrarians, but to simply speak to folks who had an open mind and to maybe surprise people along the way; to not necessarily hit every point that has been hit before, and maybe hit some that have not been hit much. Part of it is just my personality - I am intensely curious about everything. Even the most traumatic experiences in my life have largely been approached with curiosity. I want to know what is going on and understand it, and obviously there is an emotional component too that I am aware of and that I deal with, but curiosity is such a strong motivator. And I am also really a deeply felt optimist. Ultimately, the sun is going to expand and swallow this planet. That is the very long term forecast, but there is a lot to do in the meantime.

Because this topic is so complicated, because it is so potentially dire, I think that is all the more reason that we need to approach it carefully. That said, in this latest edition of my book, I am more pessimistic about any short-term traction in lowering global emissions. For various reasons, it is extremely difficult to forge a global agreement that is going to stick. There are all kinds of perverse incentives for one country or one region or one industry to increase emissions that they are saving elsewhere, partially because the cost of fuel goes down.
But I think that simply means we have to redouble our efforts to work on all levels. We have got to have global talks and global diplomacy and understanding about this issue. Industry has to respond to it. Individuals have to respond, and not have illusions that every single

action we take on a personal level is going to transform the climate, yet know that it does make a difference. It is a drop in the bucket, and it is an expression of caring about the environment, and in many cases it improves our lives. The ways we respond to climate change are often more sustainable in a human way. Taking mass transit connects you a little bit more to the people around you. These are kind of obvious things. So, in the end, I am optimistic - I would say cautiously optimistic, and hopefully optimistic - and I hope the book reflects that; it has to reflect that. If one is going to really write something that is true to oneself, then one's approach to life is going to come out in the writings.

If you were to sum up in a very pithy few phrases your approach to climate change, and how you think we all should treat this great global challenge, how would you do that?

I like to think of it in terms of the philosophy of the 12-step approach for folks who are dealing with alcohol or other issues that are addictive and compulsive. Among the things that I think are relevant from that tradition are "one day at a time: we are certainly not going to solve this problem within one day." "You change what you can and accept what you can not change." I have no illusions that we can control the entire planet, nor do I think we should. There is a lot there that is relevant. We should take it seriously, we do not give up hope or panic or see it as the end of the world.

By the same token, I really admire folks who are taking it deeply seriously and acting on that. I am not necessarily an activist myself, but I have tremendous admiration for the folks who are activists, who are especially acting from a place of knowledge and integrity and practicality. Sometimes, of course, being an activist is not always practical. You are pushing the envelope, but we need people to be pushing the envelope to move the center. You need that outer ring to get the core to move forward.

So, I do not necessarily recommend everybody being an activist, but it is important to look at what most of the activists are saying – again, folks who are acting with intelligence and reason and science — and figure out how that message speaks to you and what you can do in your own life. We all can do some things in our life. None of us can necessarily transform our entire lives, but we can do something, and something is often better than nothing. One way I think of it is in terms of having a world that is more plant-centered in terms of how we eat. I have no illusions everybody is going to become a vegetarian, but if we all skip meat for a meal or two or three a week, it will have an enormous impact on not only the welfare of animals, but literally on climate change itself, because it takes a lot more protein to feed an animal that we eat for protein than to eat plants. And for a lot of reasons, that is an attitude that can be valuable in looking at climate change.

We do what we can, and responding to a challenge can be a great adventure, whether it be bicycling more or perhaps getting an electric car if you have the resources. But it does not have to be expensive. You can simply set whatever thermostat level you are comfortable with in your house or insulate the walls where you live. Something like 40 percent or more of all the greenhouse gases in this country are because of buildings, not because of transport, and it is easy to forget that. Something as boring and unsexy as insulating your attic can really make a difference. So, I encourage people to have fun with it. Take it seriously enough to act, but not seriously enough to depress you or ruin your life.

HOLMES ROLSTON III

Holmes Rolston is widely recognized as "the father of environmental ethics" as a modern academic discipline. He is the author of seven books and is the only environmental philosopher to have lectured on all seven continents. He gave the highly prestigious Gifford Lectures, at the University of Edinburgh in 1997-1998, where he had previously received his Ph.D. He is a recipient of the Templeton Prize, past recipients of which have included Mother Teresa, Aleksandr Solzhenitsyn, the Dalai Lama, and Desmond Tutu.

Rolston's writings on environmental ethics are informed by a penetrating examination of the relevant sciences. They are imbued with a spiritual sense of wonder. He has pioneered thinking on the intrinsic value of nature, arguing that humans value nature not only for the aesthetic, industrial, recreational, and symbolic goods provided. Nature is also a good because the living beings within it have goods of their own, which humans ought to respect. Value is found at multiple levels, not only at the level of the individual but also at the genetic level and at the level of species and ecosystems as well. Most wonderful of all is the richly complex and ever varied human species itself. Yet, this appreciation for humanity need not diminish our appreciation for the "wonderland planet" upon which we live and have our being, the biospheric Earth. All of this is threatened by climate change. Each of us ought to sensitize ourselves to these multiple values of life so that we might better act to preserve this heritage for future generations.

Rolston is a founder of the journal *Environmental Ethics* and serves on the board of several other academic journals. He is a University Distinguished Professor at Colorado State University. He lives in Ft. Collins, Colorado with his wife Jane, whom he married in 1956.

THEO HORESH: *Greenhouse gas emissions are produced through everyday activities, and once released into the atmosphere spread quickly, remain there for the practical foreseeable future, and present significant harm to all life on Earth. We are thus challenged to stretch our moral commitments spatially to include all of humanity, temporally to include all foreseeable future generations, and categorically to include much of life on Earth. Each of these sorts of commitments is relatively new to moral philosophy, and each increases the burden of moral responsibility, considerably in many cases. Hence, Steven Gardner has characterized this as a "perfect moral storm," in which several moral challenges for which we are ill-prepared confront us simultaneously. And yet you have been thinking through these issues for decades. I am wondering if you can shed some light on how we might approach this perfect moral storm with wisdom and grace.*

HOLMES ROLSTON III: For many centuries humans have had to think about their families, their tribes, their churches, their governments. They have not had to think much about descendants in the distant future or people on the other side of the planet. In that sense, I think, it is new. Yet, many of these same institutions, such as democracy or the church, do invite you to think about people on the other side of the planet or children or grandchildren. Native Americans used to say if we could think for seven generations that would get what we needed. We hope we can think across more than seven generations. Most of us knew a grandparent, maybe a great-grandparent. We know children, grandchildren, maybe great-grandchildren. If you can think in that long of a time-span, then maybe you can confront some of these problems with some wisdom and some grace.

These problems are unusually complex. This will be the first time that humans have had to confront putting the planet in peril - that is new. We have confronted losing a nation. We have confronted losing a culture or a tradition. But we have not confronted putting a planet in peril. In that sense, we are at a new juncture in the history of the planet. Whether we can confront that with wisdom and grace remains to be seen.

It also seems like we are challenged to think with more mathematical precision, into the future, in a way that we have never been challenged to before. We now have cost-benefit analyses telling us what the impact of our actions is going to be five generations into the future, and this seems to make the task of thinking through these issues a lot more complex and challenging.

Yes, but you are talking to a philosopher and not an economist. So, when you begin to challenge me with numbers, I am going to respond, "Yes, but the way you interpret, the spin you put on those numbers, is not built into them." The spin you put on the numbers is going to come from a larger worldview in which you think more growth is important or in which you might think a diminished population is important. They are largely numbers in dollar signs or they may be numbers of persons. How you interpret them depends on the value system you use when you put those numbers in context. So, I am going to look over the shoulders of these people who are throwing numbers at me across generations, or into the distant future, with a pretty critical eye. You may say, "These are the numbers." And I may say, "But what are the value choices we have in dealing with these numbers?"

According to an Intergovernmental Panel on Climate Change report, there is a medium confidence that approximately 20-30 percent of all species are at risk of extinction if increases in global warming exceed 1.5 to 2.5 degrees Celsius. Say what we will about the precision or the reliability of those numbers, the numbers themselves are unbelievable. How are we to make sense of such numbers and the value of these species that are being lost?

The numbers are alarming, despite what I just said about philosophers wanting to know the framework in which the numbers go. These numbers go in the framework of extinction of species. They go in terms of climate change, and we do know something about what climate change means. So, I agree that these numbers are staggering. They are so staggering that it is hard for many people to make sense of them. We have not faced extinction in that kind of range in human history, nor have we faced climate change in the range of 1.5 to 2.5 degrees Celsius in human history, so these are alarming figures. That forces us to wonder whether humans are up to dealing with this level of complexity.

I judge that we are at some kind of hinge point in history. Given this degree of global warming and extinction of species, it looks to me as though a quick sort of response would be, "It looks like we are going to have to control our appetites; there is something self-destructing about this growth mentality." It has only been a century-and-a-half, more or less, that humans have had enormous powers for growth. For most of human history, human activities were powered by muscle and blood, human labors, also that of horses and oxen, plus a little water-power grinding grain, or wind sailing ships. It really was not until the coming of the steam engine in the mid-1800s, followed at the turn of the century by petroleum and all that came with it, that humans have had hundreds of times the power at their disposal that they once had. The use of that power is producing carbon dioxide, which is nothing anybody wanted. But like it or not, it is a result of what is happening. Now we are able to ask ourselves, "What are our options?" I am not likely to be here when all this washes out. My students are likely to be here 50 years from now – so my concern is partly for the future, but it is partly for the latter years of living people. Can we in some sense face up to what is going on?

In many ways a species is merely an abstraction. And yet we can de-

fine species very clearly and they do matter. So, how do you think about the value of a species that is being lost?

Species are real; lions are real, but the lion-lion-lion-lion-birth-death-rebirth pattern, that has been going on in Africa for at least five million years, maybe ten, is in some sense more real than the individuals. The individuals are found in a dynamic line of continuing life. In that sense, you may want to value species, the ongoing life line, more than individuals.

Many of the animal welfare people, and many other philosophers, who like to say we have to think about individuals and their lives, will get alarmed and may want to bring me up short. Only individuals are real; species are just categories we use to sort them. But I do think that groups count, communities count, nations of which we are members count, and the species lines, which are a key element in the ongoing reproduction of life, count greatly.

Over the millennia of evolution, we do not know how many species there have been. There have been somewhere from 3 to 5 billion, at least. Today on Earth there exist something in the range of at least 5 to 10 million species. By some accounts, because we do not know the insects that well, and we do not know the bacteria and so forth, we may have 100 million species alive today on Earth. That is a staggering figure that we do not want to put at threat. Do we think we humans somehow have the right to extinguish this spectacular genesis of life? We humans are here, we belong here, we are important. But maybe we are important in terms of being trustees of this creative genesis that we inherit. Maybe we are important as caretakers of this planet.

A good planet is hard to find. We have found other planets astronomically, but we do not yet know whether any of them could support life. If we did find one, planets with life are still going to be rare in the universe. But we find ourselves on a wonderland planet, which has supported life for billions of years, which continues to support it today. Americans treasure their home, mountain majesties above fruited plains. They treasure their national parks, their wilderness areas. They treasure their landscapes. Let's get all this in the picture when we are trying to count whether we want to conserve biodiversity on this wonderland planet.

The wild can be characterized as a realm of predation in which life feeds on life, and in this sense suffering and death is a natural part of the cycle of life in the wild. And yet our human impacts are causing much unnecessary suffering. What sorts of unnecessary suffering are likely to occur due to climate change, and how might we weigh the suffering of non-human life against the well-being of humanity?

You are quite correct that there is suffering in wild nature. There is predation. Animals suffer when the weather gets dry and they cannot get water. They suffer when it gets cold and they cannot find food to stay warm. The deer are not predators, but they can suffer in their natural world. So, I agree that suffering and death is part of the cycle of life, and I do not think we can or ought to eliminate it.

We can exploit nature, but what is going to be our attitude toward that? Do we care nothing about whether what we are doing to the natural world increases animal suffering? Do we care nothing for the pains of the livestock that are butchered to be on our tables to eat? We ought to be able to think about the limits of exploitation, whether our exploitation is stepping up the pain in life.

I do not want to think about how to get more and more control of all the non-human things that are out there so that we can exploit them better. I do not want to think about my life, generally, as maximizing the capacity to control others. I want to think about harmony, I want to think about community. In that ongoing life on Earth, there is going to be suffering enough. But it does not become humans to be indifferent to the additional suffering they may be causing by exploiting the Earth.

There has been a lot of talk lately about entering a new phase of life on Earth, which has been dubbed the Anthropocene. Can you talk a bit about this concept of the Anthropocene and the extent to which humanity is fundamentally altering the conditions for all life on Earth?

The Anthropocene is a dangerous term. We will have to wonder what we mean by it. The people who celebrate the Anthropocene are backing geo-engineering or re-engineering the planet pretty quickly. They say that nature is over, nature is gone, now humans are in the driver's seat. Mark Lynas says, to paraphrase, "We are the God species, nature no longer runs the Earth, we do." What we must push for, according to the Royal Society of London, the world's oldest scientific society, is *sustainable intensification* of reaping the benefits of exploiting the Earth. This idea that humans are going to manage the planet ever-more intensively for their own benefit is, I think, dangerous.

The idea that humans will ever-more cleverly engineer the planet for the next 15 thousand years strikes me as being inordinately speculative and extremely unlikely. I will take my chances with a planet that has been working more-or-less like it has been working for the last 15 thousand years. I want us to get in harmony with that and not get anxious about rebuilding the planet so that we can exploit it better or so that we can fix our problems for the next 15 thousand years. We are going to have to think more carefully about the human presence on Earth. But I want that to be thinking about harmony and community. I do not think humans want to look forward to a denatured life on a denatured planet. Basically, we want to keep life natural. Of course, we are going to farm and so forth. But we are going to require ecosystem services - the air we breathe, the water we drink, the soil we use, the ocean currents flowing. We need to think about these fundamental life support systems continuing and not being managed by arrogant human engineers.

You are somewhat unique amongst environmental philosophers in that you have placed a strong emphasis on both the value of non-human life, including species and ecosystems, and the wonder of human complexity and the genetic accomplishment of the human being. What is the significance of humanity relative to all other life, and how can we cherish what is most sacred in humanity in such a way that we gain an even-greater appreciation for the value of all other life?

That is a complex question. The last book I wrote is called, *Three Big Bangs*. Everybody knows the first big bang, the beginning of the universe some 13 or 14 billion years ago. What I call the second big bang is the explosion of life on Earth. Earth started out, once upon a time, with zero species of living things. We have already said in this conversation that we have had billions over time. We have got millions today, in terms of both increase of diversity and increase of complexity. I think the explosion of life on Earth is one of the miracles of the universe. I would be delighted if we find life elsewhere, but it is going to be rare. The third big bang is mind, right between our ears. Of course animals have minds. Elephants have minds, in some sense bigger minds than we do. But they are not more complex. Mice have minds; many creatures have brains, at least. And human brains share a lot with them.

Much of our human mind/brain is held in common with others. But there are unique capacities in the human mind that radically transcend anything known in any other species. Humans are the only species who can know they are on a planet. Humans are the only species who can form cumulative transmissible cultures; animals have only quite simple cultures. Humans can learn the Pythagorean Theorem, that was taught by Pythagoras thousands of years ago, and has been taught and retaught over centuries. We humans today can have high-technology development, but this is built on what has been transmitted from mind-to-mind-to-mind over many centuries. So, this whole conversation that we are now having about human responsibility, about human uniqueness, is just not the kind of thing we can ever imagine our dogs or chimpanzees having.

In that sense, humans are radically different. That radical difference, since we alone can know we are on a planet, we alone can know evolutionary history with its creative genesis, surely means that we ought to cherish this wonderland planet on which we live. Anything else will stunt our humanity. Now I said a minute ago, we do not want a denatured life on a denatured planet. You might want to say, "I want more civilized life." Let me put it this way: if humans do not become trustees of this planet, as we can and ought to do, it is going to stunt our humanity. If that is a self-interest, then count it as that. But you do not want to live a stunted life, do you? Then wake up to human responsibility on this wonderland planet.

Now many environmentalists would favor a world with as few as 500-million humans, and you often hear the number of a billion thrown around. But these low numbers sometimes seem to stem from a sort of misanthropy. Some environmentalists appear to hate the human experience and are not really sure what to make of the human capacity for reason. But I get the impression that you might like to see more of us around, that you value something in the achievements of human civilization.

Humans belong on the planet; humans are a marvelous species. A planet without humans would be much poorer than it is. But that does not mean the more the better. We have got seven billion now. Does that mean we shrink from seven to one billion? I don't think so – I am not wise enough to say. Does that might mean shrinking from seven to five billion? Maybe. I will just have to see when we move in that direction whether it looks like life is getting richer, better, more meaningful. I will not be around, because I am a senior citizen. But somebody will be around to wonder if this de-growth, downsizing, right-sizing, is leaving us better off. And I predict that they will find life better with a less crowded Earth.

Insofar as environmental ethics touches on the value and the interconnectedness of all life, it seems shot through with religious implications. Could you talk a little bit about the spiritual and religious implications of the rise of environmental ethics and environment philosophy in general?

You can be an environmentalist without being religious. I myself am a religious person. I think there is a fuller, richer picture than just to find yourself, surprisingly, on a wonderland Earth and wanting to save it. And what is the dimension of depth? Just go down deep enough, just think about the larger, bigger picture. We do find ourselves on a planet in which there has been a creative genesis over the millennia of natural history. And you just may say, "Well, that is a given." But others might want to say, "It is not just a given, it is a gift in some sense." There is something about these forces of creativity that have produced a marvelous Earth, and we need to think deeply about sources and origins.

The Hebrew tradition thought of the land of Israel as being a promised land. They saw the land as a gift. That can be used in mistaken ways, as a claim of privilege for a select people. But I would prefer to think that humans all over the Earth, who have lived on six of the seven continents and learned to love their landscapes, could think of all of the landscapes as being in some sense a gift, something over which we are trustees. The Hebrews had the idea that Israel was the Promised Land; it was going to flow with milk and honey. A critic might say, "That is just going to give the Israelites more milk and honey; that just sounds like they want to exploit this Promised Land. But you need to pay attention to the wider context of living in the Promised Land. The Hebrew prophets said, "the land flows with milk and honey *if and only if* 'justice flows down like waters.'" That is a line from the prophet Amos. Living well on the landscape involves living with a sense of fairness and justice. The Hebrews gave that conviction to the Christian faith. The Muslim tradition inherited and developed similar ideas.

The idea we need is a sense of humans respecting what they have been given and sharing it for a larger community of life in which there is fairness and justice. By the time you reach such thoughts, if you are not religious, you are getting are pretty close. You might say the Promised Land is just where this idea got started in Ancient Israel. Still, the idea continues forcefully. We live on a planet with promise, a marvelous wonderland planet, and the people who are most likely to think with depth about that, and to see it over long time frames, to think about sacrifices they might make to continue life on this planet, are likely to be religious or something pretty close to it. They will at least call themselves spiritual.

You have been contemplating these issues for several decades. Any wise advice for younger people trying to sort through the meaning of the changes we are now experiencing on Earth?

You do ask tough questions. Sometimes I answer that kind of question like this: my great, great-grandfather owned slaves, and we have cast off slavery, and that is a marvelous thing, at least in the United States. In my own lifetime, when I grew up in the U.S. South, black people rode at the back of buses, and they ate in separate places and restaurants; they were segregated. In my own lifetime, we have largely, if not entirely, cast off that ancient segregation.

What I am building up to is the idea that students today must not think that no big changes are possible. I am just giving you examples of two enormous changes in human history that have occurred in the last couple of lifetimes. If you had told me when I was a young man that we would have a black President of the United States, I would have said you were out of your mind. And now I celebrate the fact - whatever you think of Obama's policies - that we have at least been able to elect a black person as President of the nation. We have had black people in many high places. These are radical changes. I grew up in the U.S. South. My ancestors raised tobacco. If you told me when I was a young man that on my Colorado State University campus you would not be permitted to smoke a cigarette, or that on an airplane or a train you could not smoke a cigarette, I would have said you are out of your mind. But we have had radical changes in whether we accept smoking and whether we accept blacks and whether we accept women's rights. So, do not tell me that big changes are not possible. Do not underestimate our capacity for producing major changes in the way we live and think in our own lifetime.

AFTER PARIS

Tackling climate change may be a bit like constructing a Medieval cathedral, with each generation building on the work of previous generations. But whereas the architecture of a cathedral is planned, each generation of climate advocates will find themselves confronted by new challenges.

"It is not possible to know what is possible," as Francis Moore Lappe emphasized in this book. Things might turn out substantially better than we expected, but they might also be much, much worse. This makes climate moonshot goals tricky. It is hard to take on big long-term goals when there is a high probability the moon will radically reposition itself between the setting and achieving of the mission. Climate advocates must aim high, though, for the biosphere is in peril.

Several commentators in this book expressed skepticism over the possibility of ever passing a serious global climate deal. They reasoned that climate change is simply too big and overwhelming for global solutions. Better to chunk it down into smaller bites. Small-scale solutions are more manageable. And working to achieve them is less likely to result in disappointment and burn-out. Their words were backed by decades spent working for just such a global agreement. But sometimes the wisdom of experience fails to account for the unknown-unknowns of an ever-changing world.

As this book was coming to completion in December 2015, the dam finally broke open. The Paris Climate Deal set ambitious targets for halting the rise in temperatures, reducing deforestation, and bearing the costs of climate change. But while it built a consensus on the goals to be pursued, it provided few enforcement mechanisms. To many anxious climate watchers, 195-states coming to agreement on anything climate-related was more than they imagined possible, but to others it was too-little, too-late. Perhaps it was a bit of both.

Paris highlights several themes touched on this book. Peter Senge spoke of the need to set achievable goals and to build momentum through snowball-effects. Paris put wind in the sails of many climate advocates in this sense and will thus fuel work on climate change in every domain. Several thinkers spoke of the need to maintain positivity and to imagine something new and inspiring. And Paris was, if anything, inspiring. John Broome spoke of the need for global political action more than anything else, and Paris laid a foundation for future global climate deals. All of this occurred through the Paris Deal, so the naysayers seem to be missing a lot. And yet Paris may not have been nearly so important as many commentators have suggested.

We will need to learn to live with climate change, note both Andrew Revkin and Mike Hulme. It will be with us for a long time to come. Thus, our relationship to it will need to be made sustainable. Living with climate change will mean making changes to virtually every human institution and every aspect of our lives. The changes may be conditioned from the top-down through, say, a global carbon-tax. They may come from the bottom-up through institution building and ethical movements to decrease consumption. Either way, the changes we will have to make are enormous. Best we begin to learn to live with not just climate change but the changes in lifestyle it will necessitate.

The Paris Climate Deal lays a foundation upon which future global climate deals might be built. This is simply one-step in a long series of transformations, the most important political step perhaps, but far from the last. Building upon Paris will mean framing the issue in terms the general public can comprehend. It will mean inspiring them to action. It will mean sustaining the inspiration of people already working on the issue. It will mean continuing to innovate technologically and institutionally. And it will mean generating a willingness to transform our lives and, ultimately, human civilization.

The tasks that remain are psychological, sociological, ethical, and spiritual. They are still the most neglected among climate advocates. They still make climate change not just a curse but an opportunity. Humanity possesses in climate change the impetus needed to re-examine everything and to make the world anew. Climate change challenges us to ask just what humanity is doing on the planet, notes Mike Hulme. It is a question that will linger for centuries to come. Best we come to it sooner rather than later.

Theo Horesh,
Boulder, Colorado

Lightning Source UK Ltd.
Milton Keynes UK
UKHW020352070120
356476UK00005B/241/P